大数据丛书

"十三五"国家重点出版物出版规划项目

深入理解 Flink
实时大数据处理实践

余海峰 著

电子工业出版社
Publishing House of Electronics Industry
北京·BEIJING

内 容 简 介

本书介绍了实时数据处理引擎 Flink，讲解了流处理 API、批处理 API、机器学习引擎 FlinkML、关系型 API、复杂事件处理，以及指标度量与部署模式，分析了流式数据处理理论中时间、窗口、水印、触发器、迟到生存期之间的关联和关系，深入分析了多项式曲线拟合、分类算法、推荐算法的理论和 FlinkML 实现。

本书适合希望快速上手 Flink 以开展实时大数据处理与在线机器学习应用的从业者阅读。

未经许可，不得以任何方式复制或抄袭本书之部分或全部内容。
版权所有，侵权必究。

图书在版编目（CIP）数据

深入理解 Flink：实时大数据处理实践 / 余海峰著. —北京：电子工业出版社，2019.4
（大数据丛书）
ISBN 978-7-121-36045-9

Ⅰ. ①深… Ⅱ. ①余… Ⅲ. ①数据处理软件 Ⅳ.①TP274

中国版本图书馆 CIP 数据核字（2019）第 033826 号

策划编辑：郑柳洁
责任编辑：牛　勇
印　　刷：北京捷迅佳彩印刷有限公司
装　　订：北京捷迅佳彩印刷有限公司
出版发行：电子工业出版社
　　　　　北京市海淀区万寿路 173 信箱　　邮编：100036
开　　本：787×980　1/16　印张：19　字数：308 千字
版　　次：2019 年 4 月第 1 版
印　　次：2021 年 2 月第 7 次印刷
定　　价：89.00 元

凡所购买电子工业出版社图书有缺损问题，请向购买书店调换。若书店售缺，请与本社发行部联系，联系及邮购电话：（010）88254888，88258888。
质量投诉请发邮件至 zlts@phei.com.cn，盗版侵权举报请发邮件至 dbqq@phei.com.cn。
本书咨询联系方式：010-51260888-819，faq@phei.com.cn。

前言

实时大数据是与时俱进的变革

从互联网时代的数据爆炸,到即将大规模铺开的 5G 通信支撑下的物联网时代的大数据浩海,作为赋能工具的大规模数据处理,技术架构起到了决定性的作用,反过来也推动了技术架构与时俱进。

在谷歌公司发表的三篇划时代论文(分别介绍 MapReduce、GFS 和 BigTable)的推动下,开源项目 Hadoop 横空出世,并于 2008 年 1 月正式成为 Apache 的顶级项目;此后,Hadoop 迅速建立起大数据生态体系,并由此衍生出一系列大数据处理的理论和与之对应的大数据处理框架:从批处理到流处理,从 Hadoop 到 Storm / Spark,再到 Flink。本书将阐述大数据实时处理理论的变迁,并着重介绍流处理框架 Flink。

数据处理任务往往需要对全量数据进行计算,而全量数据很难使用传统关系型数据库进行批量计算,原因如下:

(1)磁盘寻址时间的提升速度远远落后于磁盘带宽的提升速度。如果数据访问包含大量的磁盘寻址,则大数据处理势必带来较大的延迟,因此基于传输带宽设计大数据处理系统更符合现状。

(2)相比全量数据计算,关系型数据库适用于在线事务处理(OLTP,On-Line Transaction Processing)场景,查询和更新是其设计的要点,索引是主要的设计方案。但是在大数据集的场景下,索引的效率往往不如全量扫描。因此,Hadoop 应运而生,借助 MapReduce 计算引擎成功解决了大数据所面临的可计算(可参考谷

歌的论文 *MapReduce:Simplified Data Processing on Large Clusters*)、伸缩、容错等困难，成为大数据系统的标配组件。

数据爆炸式增长，以及数据处理的实时性要求越来越高，大数据处理系统越来越复杂。在这种情况下，传统的 OLTP+OLAP（On-Line Analysis Processing，在线分析处理）系统架构不堪重负：复杂的数据需求实现流程、过载的数据仓库抽取任务队列、不同的技术栈带来的需求理解偏差等将导致数据从 IT 部门到 DT 部门的周期过长；微服务方法的大规模应用，导致在分布式系统中维护全局状态的一致性异常困难，而以数据流作为中心数据源的流处理方法能有效规避这种困难。有的学者甚至提出通过合理的架构设计，打破 CAP 定理。因此，低延迟、强一致性、适用于乱序的流处理框架 Flink 正席卷而来，即将成为大数据领域流处理的标配组件。

本书特色

本书将从多个角度讲解同一个技术概念：

（1）分析引入 Flink 这个技术概念的原因，使读者能够快速理解相关技术的应用场景，如为什么需要实时数据处理、为什么需要机器学习架构、为什么需要关系型 API、为什么需要复杂事件处理。

（2）剖析 Flink 技术的理论创新过程，使读者能够深入理解 Flink 的理论基础，使 Flink 应用开发架构师或工程师能够游刃有余地解决线上系统遇到的实际问题，如 Flink 一致性保证的异步检查点屏障的理论创新过程、机器学习中分类和推荐算法的分布式实现的理论创新过程、复杂事件处理的自动机理论创新。

（3）解析 Flink 编程 API 的架构。使读者可以从理论框架与 Flink 架构实现两个角度体会这个技术概念的内涵。

（4）总结应用 API 编程解决实际问题的方法。使读者能够在理解理论和编程 API 的基础上编程解决实际问题。

（5）在每章的最后一节提出一些开放式的思考题，以便读者思考这些技术概念之间的关联。

内容组织概要

本书共分为 8 章，每章的基本内容概要如下。

第 1 章流式数据架构理论　首先，概述大数据处理架构的演进历程，使读者了解大数据处理架构正在经历怎样的变革。以韩国 SK 电信的 Driving Score 技术架构演变和流式数据架构在机器学习领域的应用为例，介绍流式数据架构的产生背景及应用场景；其次，梳理流式数据架构中主要概念间的关联和关系，并以实例分析根据事件时间开滚动窗口的内在机制；最后，论述流式数据架构中一致性理论的基础及实现方式。

第 2 章编程基础　首先，概述 Flink 的特征，使读者快速了解 Flink 是什么；其次，搭建流处理和批处理 IDEA 开发环境，使读者感性地了解 Flink 是怎么解决数据处理问题的；最后，介绍 Flink 的编程模型与运行时结构，如 API 分层关系、应用程序基本结构、运行时结构、任务调度和物理执行计划。

第 3 章流处理 API　首先，概述流处理 Pipeline、Source 和 Sink 的常见形式；其次，介绍时间特征设置与水印生成器、算子、窗口机制、连接器、状态管理与检查点编程。

第 4 章批处理 API　首先，概述批处理的程序结构，Source、Sink 与连接器的常见形式，以及常见的算子函数；其次，以两个机器学习的例子来介绍迭代的应用；最后，介绍批处理程序优化的语义注解形式。

第 5 章机器学习引擎架构与应用编程　首先，在总结 Scikit-learn 架构实践的基础上，详细分析 FlinkML 的底层实现代码；其次，分析多项式曲线拟合、分类算法、推荐算法的理论和代码实现。本章有大量代码分析，是流处理与批处理 API 编程的实战部分。

第 6 章关系型 API　首先，介绍 Flink 引入关系型 API 的原因，以及 SQL 解析与优化框架 Calcite；其次，详述关系型 API 的主要内容；最后，介绍架构在 DataStream 上的关系型 API 的背后机制，即动态表。

第 7 章复杂事件处理　首先，以股票异常交易检测为例讲述模式匹配的编程过程，以及流处理 API 和关系型 API 在解决这类问题时遇到的困难；其次，介绍 NFA[b] 模式匹配编程模型；最后，基于 Flink CEP API 编程解决股票异常交易检测问题。

第 8 章监控与部署　讲述指标度量的编程模型和 Flink 集群部署模式。

联系作者

购买本书的同时，你将获得免费向作者寻求帮助的权利。每章最后一节为思考题，设置思考题的目的是加深读者对本书的理解。这些问题是开放性的，因此书中并没有给出标准答案，读者可以通过作者的微信公众号（见前勒口）获取帮助。

读者服务

轻松注册成为博文视点社区用户（www.broadview.com.cn），扫码直达本书页面。

- **提交勘误**：您对书中内容的修改意见可在【提交勘误】处提交，若被采纳，将获赠博文视点社区积分（在您购买电子书时，积分可用来抵扣相应金额）。
- **与读者交流**：在页面下方【读者评论】处留下您的疑问或观点，与其他读者一同学习交流。

页面入口：http://www.broadview.com.cn/36045

目录

第 1 章 流式数据架构理论 1
1.1 大数据处理架构演进历程 1
1.2 案例分析 8
1.2.1 SK 电信驾驶安全性评分 8
1.2.2 流式机器学习应用 12
1.3 流式数据架构基本概念 17
1.3.1 流 17
1.3.2 时间 18
1.3.3 窗口 21
1.3.4 水印 23
1.3.5 触发器 23
1.3.6 数据处理模式 23
1.3.7 如何理解流式数据架构的内在机制 27
1.4 根据事件时间开滚动窗口 28
1.4.1 what：转换 / where：窗口 29
1.4.2 when：水印 29
1.4.3 when：触发器 32
1.4.4 when：迟到生存期 34
1.4.5 how：累加模式 35
1.5 一致性 37

1.5.1　有状态计算 37
　　　1.5.2　exactly-once 语义 38
　　　1.5.3　异步屏障快照 39
　　　1.5.4　保存点 44
　1.6　思考题 45

第 2 章　编程基础 46

　2.1　Flink 概述 46
　2.2　让轮子转起来 47
　　　2.2.1　本书约定 47
　　　2.2.2　搭建单机版环境 48
　　　2.2.3　配置 IDEA 51
　2.3　编程模型 53
　　　2.3.1　分层组件栈 53
　　　2.3.2　流式计算模型 54
　　　2.3.3　流处理编程 57
　2.4　运行时 62
　　　2.4.1　运行时结构 62
　　　2.4.2　任务调度 66
　　　2.4.3　物理执行计划 69
　2.5　思考题 70

第 3 章　流处理 API 71

　3.1　流处理 API 概述 71
　3.2　时间处理 73
　　　3.2.1　时间 73
　　　3.2.2　水印 74
　　　3.2.3　周期性水印生成器 75
　　　3.2.4　间歇性水印生成器 77
　　　3.2.5　递增式水印生成器 78
　3.3　算子 79
　　　3.3.1　算子函数 80

3.3.2	数据分区	83
3.3.3	资源共享	85
3.3.4	RichFunction	85
3.3.5	输出带外数据	86

3.4 窗口 86
  3.4.1 窗口分类 87
  3.4.2 窗口函数 90
  3.4.3 触发器 94
  3.4.4 清除器 96
  3.4.5 迟到生存期 96

3.5 连接器 97
  3.5.1 HDFS 连接器 98
  3.5.2 Kafka 99
  3.5.3 异步 I/O 102

3.6 状态管理 104
  3.6.1 状态分类 104
  3.6.2 托管的 Keyed State 104
  3.6.3 状态后端配置 106

3.7 检查点 107
3.8 思考题 108

第 4 章 批处理 API 109

4.1 批处理 API 概述 109
  4.1.1 程序结构 110
  4.1.2 Source 111
  4.1.3 Sink 112
  4.1.4 连接器 112

4.2 算子 113
  4.2.1 算子函数 113
  4.2.2 广播变量 121
  4.2.3 文件缓存 122
  4.2.4 容错 123

4.3 迭代 123
4.3.1 深度神经网络训练 123
4.3.2 网络社团发现算法 125
4.3.3 Bulk Iteration 127
4.3.4 Delta Iteration 的迭代形式 128
4.4 注解 130
4.4.1 直接转发 130
4.4.2 非直接转发 131
4.4.3 触达 132
4.5 思考题 132

第 5 章 机器学习引擎架构与应用编程 133
5.1 概述 133
5.1.1 数据加载 134
5.1.2 多项式曲线拟合的例子 135
5.2 流水线 137
5.2.1 机器学习面临的架构问题 137
5.2.2 Scikit-learn 架构实践总结 138
5.2.3 FlinkML 实现 140
5.3 深入分析多项式曲线拟合 170
5.3.1 数值计算的底层框架 170
5.3.2 向量 172
5.3.3 数据预处理 178
5.3.4 特征变换 184
5.3.5 线性拟合 188
5.4 分类算法 190
5.4.1 最优超平面 190
5.4.2 凸优化理论 193
5.4.3 求解最优超平面 198
5.4.4 核方法 200
5.4.5 软间隔 205
5.4.6 优化解法 208

- 5.4.7 SVM 的 FlinkML 实现 211
- 5.4.8 SVM 的应用 220
- 5.5 推荐算法 221
  - 5.5.1 推荐系统的分类 221
  - 5.5.2 ALS-WR 算法 223
  - 5.5.3 FlinkML 实现 225
  - 5.5.4 ALS-WR 的应用 230
- 5.6 思考题 230

第 6 章 关系型 API 234
- 6.1 为什么需要关系型 API 234
- 6.2 Calcite 235
- 6.3 关系型 API 概述 236
  - 6.3.1 程序结构 236
  - 6.3.2 Table 运行时 239
  - 6.3.3 表注册 241
  - 6.3.4 TableSource 与 TableSink 242
  - 6.3.5 查询 244
  - 6.3.6 相互转换 244
- 6.4 动态表概述 247
  - 6.4.1 流式关系代数 247
  - 6.4.2 动态表 248
  - 6.4.3 持续查询 250
- 6.5 思考题 255

第 7 章 复杂事件处理 256
- 7.1 什么是复杂事件处理 256
  - 7.1.1 股票异常交易检测 256
  - 7.1.2 重新审视 DataStream 与 Table API 258
- 7.2 复杂事件处理的自动机理论 259
  - 7.2.1 有穷自动机模型 NFA 259
  - 7.2.2 NFA[b] 模型 261

7.2.3　带版本号的共享缓存 263
7.3　FlinkCEP API 265
　　7.3.1　基本模式 266
　　7.3.2　模式拼合 267
　　7.3.3　模式分组 268
　　7.3.4　匹配输出 269
7.4　基于 FlinkCEP 的股票异常交易检测的实现 270
7.5　思考题 274

第 8 章　监控与部署

8.1　监控 275
　　8.1.1　度量指标 275
　　8.1.2　指标的作用域 279
　　8.1.3　监控配置 279
8.2　集群部署模式 281
　　8.2.1　Standalone 281
　　8.2.2　YARN 281
　　8.2.3　高可用 284
8.3　访问安全 284
8.4　思考题 286

参考资料 287

第 1 章
流式数据架构理论

在移动互联网领域，个性化服务、极致的用户体验要求业务系统具备实时数据处理能力，传统的批处理数据架构已经不堪重负。经过一系列理论创新与实践探索，流式数据架构 Flink 在实时数据处理领域取得了巨大成功，正成为大数据处理的标配框架。

为了让读者厘清大数据处理架构变革的源与流，1.1 节先概述大数据处理架构的演进历程，如 Storm、Spark、Lambda、Flink；为了让读者更容易理解流式数据架构思想，1.2 节将以韩国 SK 电信的 Driving score 技术架构演变和流式数据架构在机器学习领域的应用为例，介绍流式数据架构的产生背景及应用场景；1.3 节将介绍流、时间、窗口、水印、触发器等，并在这些概念的基础上剖析数据处理的各种模式；1.4 节将梳理流式数据架构中主要概念间的关联和关系，并以实例分析根据事件时间开滚动窗口的内在机制；1.5 节将论述流式数据架构中一致性理论的基础及实现方式，如有状态计算、检查点、保存点等概念。

1.1 大数据处理架构演进历程

谷歌发表的三篇划时代论文（分别介绍 MapReduce、GFS 和 BigTable），特别是介绍 MapReduce 的那篇论文，开启了大规模数据处理波澜壮阔的发展历程。一篇篇论文和那些大数据从业者耳熟能详的大数据处理架构，是这个历程中的重要里程碑，图 1-1 所示为主流大数据处理架构的发展历程。

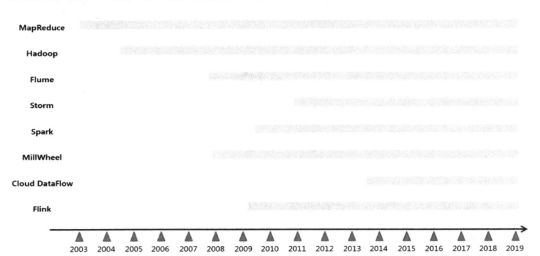

图 1-1　主流大数据处理架构的发展历程

2003 年，谷歌的工程师便开始构建各种定制化数据处理系统，以解决大规模数据处理的几大难题：**大规模数据处理特别困难**（Data Processing is hard），这里的难有多个方面，仅仅是在大规模数据上构建一个处理程序也不是一件容易的事情；保证**可伸缩性很难**（Scalability is hard），让处理程序在不同规模的集群上运行，或者更进一步，让程序根据计算资源状况自动调度执行，也不是一件容易的事情；**容错很难**（Fault-tolerance is hard），让处理程序在由廉价机器组成的集群上可靠地运行，更不是一件容易的事情。这些困难促使 MapReduce（不是 Hadoop 中的 MapReduce）诞生。MapReduce 将处理抽象成 Map+Shuffle+Reduce 的过程，这种抽象对大数据处理理论变革有着深远的影响。

以计算词频为例，MapReduce 将输入（Input）文本以行为单位分片（Split），每个 Map 任务将分片中的每个词映射为键值对的形式（Dear, 1），Shuffle 将相同键的记录组合在一起，最后由 Reduce 任务计算词频并输出（Output）结果，图 1-2 描述了一个有 3 个 Map 和 3 个 Reduce 的词频计算过程。

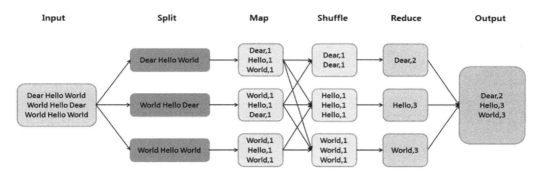

图 1-2 基于 MapReduce 计算词频的过程

笔者有一段相似的架构经历，能够帮助读者更好地理解是什么驱动谷歌的工程师开发 MapReduce 这个通用框架。驱动笔者开发一个定制化数据处理程序的想法主要来自业务需求，也有 MapReduce 思想的启发。当时，笔者就职的公司有 TB 级的短文本数据，笔者需要将这些文本的一些相邻行合并成一条记录，再对这些记录进行聚合操作，并在这之上构建一个用于语义分析的应用。出于保密要求，这些数据被**分批**归集到公司内网的一台 x86 服务器上，语义分析程序也运行在这台内网机器上。笔者有两套方案，其中一套方案使用 Hadoop，但是由于只有两台物理机器，而且用 Hadoop 有点"大炮打蚊子"的感觉，加之因着迷于 Linux 内核之美而"继承"下的"一言不合便动手造轮子"的理念，笔者决定采用另一套方案：使用 Java 语言自己动手构建一个简易的、定制化的多线程数据处理框架（类 MapReduce 数据处理框架），如图 1-3 所示。

其中，Reader 用于并行读取数据；Dealer 用于实现可级联的数据处理逻辑，如先计算记录总数，再过滤非目标记录，最后分词并计算语义标签；Writer 将 Dealer 处理的最终结果以配置的格式写入输出文件。

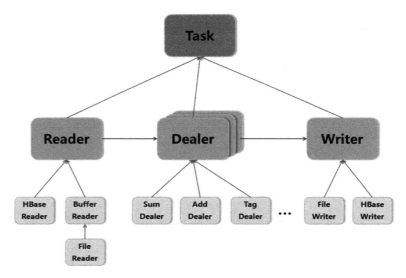

图 1-3 类 MapReduce 数据处理框架

多线程并行处理将程序运行速度提高了好几个量级。尽管如此，这段经历也令笔者回味深长：

（1）语义分析应用程序和底层组件间耦合得太紧，以至于这套软件只能由笔者维护。因为承担这个任务的部门的其他同事都是做数据分析的，没有软件开发工作经验。

（2）语义分析训练通常是相当耗时的，没有功能更强大的框架支持，手工操作的时间成本比较高。

这段经历让笔者深刻领悟到 MapReduce 框架的深思熟虑。

2004 年，Doug Cutting 和 Mike Cafarella 在构建 Nutch 时受到谷歌公司发表的 MapReduce 论文的启发，实现了开源版本的 MapReduce，即 Hadoop。此后，Pig、Hive、HBase 等工具不断涌现，Hadoop **批处理**生态系统蓬勃发展，也让人们再次领教了开源的力量，图 1-4 展示了 Hadoop 生态系统。

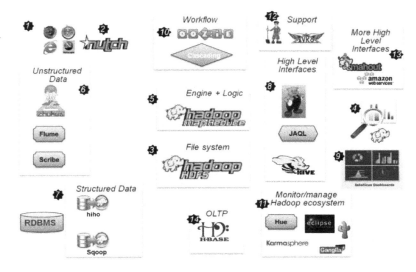

图 1-4　Hadoop 生态系统

批处理（batch）的概念由来已久。在操作系统理论中，批处理是指用户将一批作业提交给操作系统后就不再干预，由操作系统控制它们自动运行。这种操作系统被称为批处理操作系统，它是为了提高 CPU 的利用率而提出的一种操作系统。例如，在 DOS 和 Windows 系统中，我们可以在扩展名为 .bat 的脚本文件中顺序定义一系列操作命令，让操作系统自动运行这些命令。

在数据处理理论中有对应的批处理系统。批处理系统的核心功能是在大容量静态数据集上运行预定义功能的、可预期完成时间的计算任务。这里的静态是指数据集是有界的，是数据集的时间属性。

流处理（streaming）系统则是构建在无界数据集之上的、可提供实时计算的另一类数据处理系统。

经过一段时间的应用实践，MapReduce 的缺陷也逐渐暴露，最让人诟病的是 Map+Shuffle+Reduce 编程模型导致计算作业效率低下。为此，2007 年，谷歌发起了 Flume 项目。起初，Flume 只有 Java 版本，因此也被称为 Flume Java（这里所说的 Flume 和 Apache 的 Flume 不同）。Flume 将数据处理过程抽象成计算图（有向无环图），数据处理逻辑被编译成 Map+Shuffle+Reduce 的组合，并加入物理执行计划优化，而不是简单地将 Map+Shuffle+Reduce 串联。

Flume 引入的管道（Pipeline）、动态负载均衡（谷歌内部称为液态分片）和流语义思想成为大数据处理技术变革的宝贵理论财富。

产生于处理推特信息流的流式数据处理框架 Storm 以牺牲强一致性换取实时性，并在一些场景下取得了成功。为了让数据处理程序兼备强一致性和实时性，工程师们将强实时性的 Storm 和强一致性的 Hadoop 批处理系统融合在一起，即 Lambda 架构。其中，Storm 负责实时生成近似结果，Hadoop 负责计算最终精准结果。Lambda 架构需要部署两套队列集群，数据要持久化存放两份，这会导致数据冗余，增加系统维护成本。Lambda 架构示意图，如图 1-5 所示。

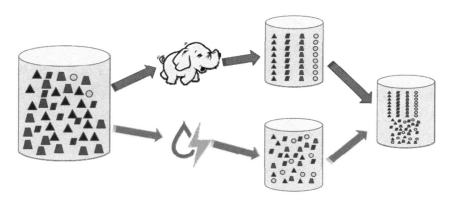

图 1-5　Lambda 架构示意图

MapReduce 模型严重依赖分布式文件系统，如 Map 将计算结果临时写入文件系统，而 Shuffle 从文件系统中读入该结果，这往往会产生较大的计算性能损耗，因此基于内存的计算是另一个选择，这就是 Spark 成功的秘诀。此外，Spark 还支持流式数据处理，即 Spark Streaming，其原理是将多个微批处理任务串接起来构建流式数据处理任务。但是这种采用微批重复运行的机制牺牲了低延迟和高吞吐的优势，引发了 Spark Streaming 是不是真正流式数据处理引擎的争议。Spark

Streaming 流式数据处理任务的架构方案,如图 1-6 所示。

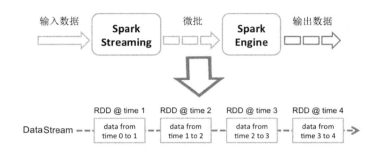

图 1-6 Spark Streaming 流式数据处理任务的架构方案

这期间,流式数据处理继续发展,出现了 MillWheel、Kafka 和 DataFlow。exactly-once 语义、数据源端的持久化和可重放、动态表理论,以及时间、窗口、水印、触发器等流式数据处理核心理论的提出,加快了流式数据处理框架的发展步伐。

作为流式数据处理中容错的解决方案之一的轻量级快照机制,借助上述流式数据处理相关理论,以及开源的旺盛生命力,Flink 于 2015 年迅速登上实时数据处理的舞台,并将推动大数据发展新的浪潮。正是在这种背景下,笔者决定深入 Flink 实现底层,为读者呈现其中的智慧之光。Flink 架构,如图 1-7 所示。

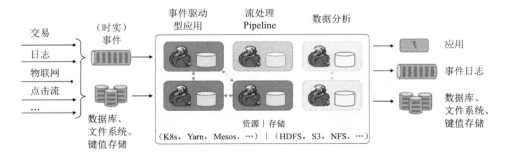

图 1-7 Flink 架构

1.2 案例分析

1.2.1 SK 电信驾驶安全性评分

SK 电信是韩国的移动通信运营商，T map 是其手机导航 App，类似于我国的百度地图和高德地图。这款 App 可对用户的驾驶安全性（超速、加速、减速）评分，汽车保险公司根据这个评分计算保费。T map 的 Driving score 功能，如图 1-8 所示。

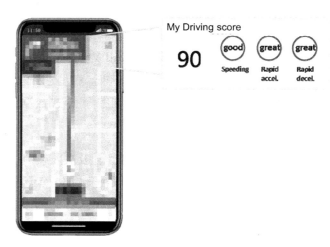

图 1-8　T map 的 Driving score 功能

1. 批处理架构

在行驶过程中，车辆的 GPS 位置（经度、纬度、海拔）信息由 App 实时上传至后台服务端。其中，GPS 一分钟定位一次，App 将五分钟内的位置信息打成一个数据包上传。

最初，系统采用批处理架构，GPS 位置信息被定期抽取到 Hive 数据仓库中，计算 Driving score 的 ETL（Extract Transform Load）批处理程序由 Oozie 调度执行，频率为每天两次，评分结果在第二天返回用户。从传统金融领域到电信领域，从零售领域到物联网领域，这种数据处理系统已经成为标配架构。但是这种 **T+1**

处理架构的缺陷也很明显,即用户体验较差、决策反应速度较慢。Driving score 批处理系统的架构如图 1-9 所示。

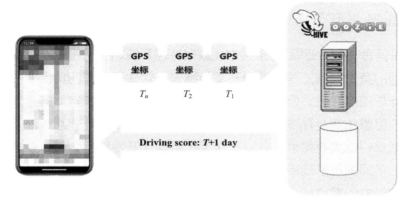

图 1-9　Driving score 批处理系统的架构

因此,SK 电信将上述批处理架构重构为基于消息的流式数据处理架构,提供实时 Driving score 服务。

2. 流式数据处理架构

笔者以 12 分钟一趟的驾驶为例,分析流式数据处理架构设计面临的问题及其解决方案。

App 端将每 5 分钟内的 300 个 GPS 位置信息以 JSON 格式打包发给后台服务端,按照定位时间先后顺序,以 a、b、c、d 命名这 4 个数据包,系统在接收数据包 d 后立即返回 Driving score。在计算机术语里,这 12 分钟一趟的驾驶称为 session(一次会话)。针对这种应用场景做如下分析。

(1) 乱序是常见情况:数据包 d 不仅包含 GPS 位置信息,还包含 session 结束标志信息。即使系统能保证数据包在 App 端的上传顺序,也不能保证数据包到达处理节点的顺序,因为后台服务系统的分布式特性可能会导致数据包乱序,如计算 Driving score 的处理程序可能先收到数据包 d,隔一段时间才收到数据包 c。

（2）窗口机制：在很多应用系统中，开始与结束标志信息并不存在，数据处理程序需要根据应用系统的领域知识推定。因此，什么时间点计算 Driving score 是流式数据处理架构设计的关键点，这个关键点在流式数据处理架构理论中被称为窗口机制。

（3）松耦合：在企业级 IT 架构中，数据处理系统和业务逻辑系统通常是分开的，数据部门负责数据处理系统的开发与维护，软件开发部门负责业务逻辑系统架构的设计与程序开发。但是，由于数据处理系统的数据来自业务逻辑系统，这两个系统不可避免地会存在一定的耦合。并且，如果数据处理系统涉及底层软件模块，数据部门就需要将相关开发工作委托给软件开发部门，但复杂的数据需求实现流程、不同的技术栈带来的需求理解偏差等因素，将导致数据从软件开发部门到数据部门的周期过长，这是另一种耦合，因此流式数据处理架构需要松耦合设计。

基于以上分析，流式数据处理系统的前端部署分布式消息系统 Kafka，直接串接在 GPS 位置信息的传输通道上，Source 节点从 Kafka 订阅 GPS 位置信息，并将 JSON 格式解析后的信息推送到时间窗口节点（Time & Window）处理，Sink 节点将处理的结果信息写入 Kafka，Kafka 负责持久化或对接到下游系统。Driving score 流式架构的逻辑模式如图 1-10 所示。

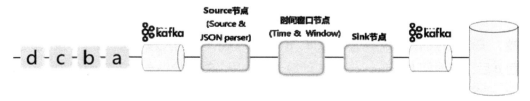

图 1-10 Driving score 流式架构的逻辑模式

在图 1-10 中，每个处理节点只有一个实例，这是流式数据处理系统的逻辑模式。流式数据处理系统是分布式计算系统，因此在实际执行过程中每个相同功能的节点都会有多个并行实例对应于物理部署模式，如图 1-11 所示。

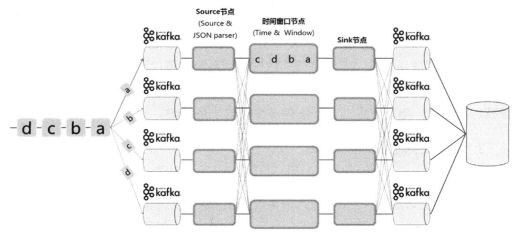

图 1-11 Driving score 流式架构的物理部署模式

在物理部署模式中，每个 Source 节点实例会和所有时间窗口节点实例相连，这样同一个 session 的 GPS 位置信息可以被同一个时间窗口节点实例处理。数据处理"一直"在运行，而不是定时运行的（批处理采用定时触发运行机制），从而规避了 $T+1$ 处理架构的缺陷。

在物理部署模式中，我们还观察到 4 个数据包的顺序为 a、b、d、c，而不是按事件发生的顺序 a、b、c、d。下面介绍解决上述数据包乱序问题的三种窗口机制。

1. 设定固定时间间隔的 session 窗口

在时间轴上，两趟驾驶不会出现重叠，即两个活动 session 之间会有一定的时间间隔，可以通过数据分析的方法计算这个间隔的合理值（不一定是最小值，可参考 1.4 节和 1.5 节的内容），例如 1 小时。如果时间窗口节点实例在接收 d 之后的 1 小时内没有再观察到新的 GPS 位置信息数据包，则系统可以在这个时间点上返回 Driving score。这种窗口机制很简单，但是仍需要等待 1 小时才能得到结果。

2. 设定 session 的事件推进标志

流式数据处理架构使用水印来推进事件时间，即 Source 节点或业务逻辑系统在 GPS 位置信息流中定期插入时间推进控制信息，用于表明某个时间点之前的 GPS 位置信息数据包均已接收完毕，这样时间窗口节点能够**实时**计算 Driving score。从这层意义上看，水印是流式数据处理系统中事件流转的同步信号。虽然这种窗口机制的用户体验好，但是需要深刻理解应用领域知识，生成水印的代价较高，特别是，当企业级业务系统有多个关联子系统时，设计水印异常复杂。

3. 触发器实时生成近似结果

对数据部门而言，深刻理解应用领域知识，特别是深刻理解应用系统架构知识的代价较高。为此，流式数据处理系统提供触发器机制以实时生成近似结果，即数据处理系统行先计算出正常的 session 结束时间点，并在这个时间点上实时计算出 Driving score，如果之后观察到属于这个 session 的迟到的 GPS 位置信息数据包，那么时间窗口节点会撤回之前的结果并生成新的 Driving score。这种窗口机制广泛应用在一些对实时性要求比较高的数据处理场景中。

1.2.2 流式机器学习应用

1. 线下训练模型、线上实时抽取特征

在机器学习中，求得一个与训练数据集相吻合的函数的过程被称为数据拟合，也被称为曲线拟合。下面是用 Python 语言编写的多项式拟合程序：

```python
import matplotlib.pyplot as plt
import pandas as pd
from statsmodels.tsa.stattools import adfuller
import statsmodels.tsa.stattools as st
import numpy as np
import math

from sklearn.linear_model import LinearRegression
from sklearn.preprocessing import PolynomialFeatures

#根据特征训练模型
    #实例化一个八次多项式特征实例
    quadratic_featurizer = PolynomialFeatures(degree=8)
```

```
    #用多项式对样本 x 的值做变换
    Feature = quadratic_featurizer.fit_transform(train_x)
    #创建一个线性回归实例
    Model = LinearRegression()
    #以多项式变换后的样本 x 的值为输入，代入线性回归模型做训练
    Model.fit(Feature, train_y)
#计算模型的拟合效果
    train_y_p = regressor_quadratic.predict(quadratic_featurizer.transform(train_x))
    train_y_plus = train_y_p - train_y
    train_y_s = pd.Series([train_y_plus[i][0] for i in
np.arange(0,int(len(train_y_plus)))])
    #标准差
    train_y_s_std = train_y_s.values.std()
    #拟合优度
    score = regressor_quadratic.score(X_train_quadratic, train_y)

#根据 x 的值预测 y 值
    y = Model.predict(quadratic_featurizer.transform(x))
```

这种机器学习算法根据训练数据集（train_x,train_y）的八次多项式特征（Feature）训练线性回归模型（Model）。在模型应用于线上预测时，流式处理系统从实时数据（x）中抽取特征（quadratic_featurizer.transform(x)），模型根据抽取的特征输出预测值（y）。

这种以 Feature 和 Model 为核心组件，线下训练模型、线上实时抽取特征的机器学习方法在传统金融安全等领域有着广泛的应用。其中，反欺诈是金融机构 IT 团队面临的主要挑战之一。欺诈的形式从 App 签名漏洞、ATM 盗刷分离器、网络仿冒到恶意商业间谍 APT（Advanced Persistent Threat，如"银行大盗" Carbanak 木马可命令受感染的 ATM 直接吐钱），欺诈的规模从个体犯罪逐步演变成有组织的犯罪。为了应对不断升级的欺诈手段，金融机构的应对措施也从最开始的手动检测、规则检测，发展到基于监督学习的模型检测、基于非监督学习的异常检测。

考虑到安全形势的严峻性、金融业务的复杂性及分支机构的跨地域性，反欺诈系统可以采用线下训练模型、线上实时抽取特征的机器学习方法。为了便于共享与分发，通常采用 PMML 的方式定义线上机器学习模型的需求，如图 1-12 所示。

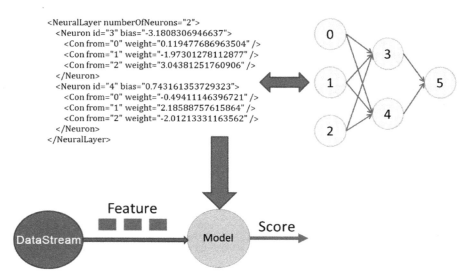

图 1-12 采用 PMML 的方式定义线上机器学习模型的需求

PMML（Predictive Model Markup Language，预测模型标记语言）利用 XML 描述和存储数据挖掘模型，是一种在不同应用程序之间快速共享模型的可选方案，已经被 W3C 所接受。PMML 中定义了数据挖掘不同阶段的相关信息：头信息（Header）、数据字典（Data Dictionary）、挖掘模式（Mining Schema）、数据转换（Transformation）、模型定义（Model Definition）和评分结果（Score Result）。

2. 实时机器学习

随着电子商务的高速发展，互联网电商平台的数据呈爆炸式增长，其存量数据可达 EB（Exabyte，艾字节，1EB=1024PB）级，每天增量数据可达 PB（Petabyte，1PB=1024TB）级，这会带来巨大的模型训练计算量。

反向传播算法（详细分析参见第 4 章）是深度学习模型训练的通用算法，如果进行全量样本训练，则每次训练需要计算所有样本的损失和，再进行梯度运算，计算开销太大，很难进行在线学习；如果进行单样本训练，那么每次训练只计算每个样本的损失与梯度,但是整个模型训练的收敛速度和泛化效果都不理想。

我们先将样本数据分成多个**微批**(microBatch)数据集,然后在每个微批数据集上进行梯度运算。下面的代码片段是使用 TensorFlow 实现 Word2Vec 的例子:

```
// 定义 Word2Vec 训练网络有向计算图
graph = tf.Graph()
    with graph.as_default():
        ...
        loss = tf.reduce_mean(tf.nn.nce_loss(weights=nce_weights,
                                              biases=nce_bias,
                                              labels=train_labels,
                                              inputs=embed,
                                              num_sampled=num_sampled,
                                              num_classes=vocabulary_size
                                              ))
        optimizer = tf.train.GradientDescentOptimizer(1.0).minimize(loss)

        L2_norm = tf.sqrt(tf.reduce_sum(tf.square(embeddings), 1 , keep_dims=True))
        normalized_embeddings = embeddings / L2_norm
        valid_embeddings = tf.nn.embedding_lookup(normalized_embeddings,
valid_dataset)
        similarity = tf.matmul(valid_embeddings, normalized_embeddings,
transpose_b=True)
        init = tf.global_variables_initializer()
    ...
    with tf.Session(graph=graph) as session:
        init.run()
        ...
        // 微批训练
        for step in range(num_steps):
            ...
            feed_dict = {train_inputs : microBatch_inputs, train_labels :
microBatch_labels}
            _,loss_val = session.run([optimizer, loss], feed_dict=feed_dict)
            ...
```

此外,电商数据的特征并不平稳,如"双 11"时商品的价格、活动的规则与平时完全不同,通过历史样本数据训练出的模型往往不能拟合出让人满意的结果,图 1-13 概括了这种非实时机器学习方式的特征。

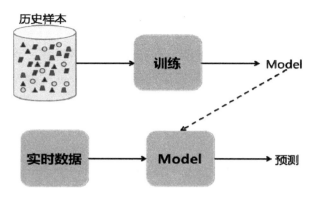

图 1-13 非实时机器学习方式的特征

实时机器学习是电商应用的合理选择：线上实时特征抽取，采用微批方式实时训练模型。图 1-14 展示了阿里巴巴实时机器学习模型的架构。

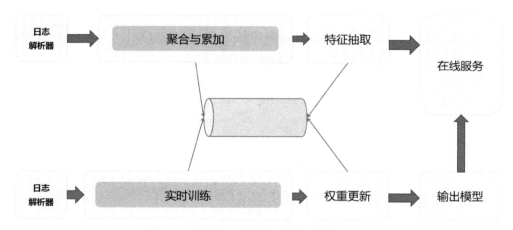

图 1-14 阿里巴巴实时机器学习模型的架构

图 1-14 中的上半部分为实时抽取数据特征的过程，下半部分为实时训练的过程，这样整个系统可以根据实时特征更新模型参数，并将更新后的模型部署到线上。

1.3 流式数据架构基本概念

1.3.1 流

流往往代表事物的无尽特征,它丰富的内涵往往会造成人们对流式数据处理是什么的误解,以及对流式数据处理系统特征(能做什么、不能做什么)的混淆。在开始详述流式数据处理理论之前,我们先定义流。

最初,流式数据处理是通过批处理系统实现的,如 Spark Streaming,其原理是将多个微批处理任务串接起来构建流式数据处理任务。但是这种采用微批重复运行的机制,牺牲了延迟和吞吐,引发了 Spark Streaming 是不是真正流式数据处理引擎的争议。为此,业界便开始构建用于处理没有时间边界数据(无界数据集,Unbounded Data)的实时数据系统。

因此,从这个角度可以定义流是一种为无界数据集设计的数据处理引擎,这种引擎具备以下特征:

(1)具备强一致性,即支持 exactly-once 语义。

(2)提供丰富的时间工具,如事件时间、处理时间、窗口。

(3)保证系统具有可弹性、伸缩性。

(4)同时保证高吞吐、低延迟与容错。

(5)支持高层语义,如流式关系型 API(SQL)、复杂事件处理(CEP,Complex Event Processing)。

此外,本书中还用流式数据处理系统、流式数据处理架构及流处理等具有相同内涵的术语指代 streaming,其中流式数据处理架构用在描述系统架构的语境中。虽然"流处理"这个词用得较为广泛,但是本书只在和批处理对比分析的语境中使用流处理这个口语化称呼。

1.3.2 时间

在流式数据处理理论中，经常使用事件或记录表示从所处理的数据集中拉取的数据，在 Flink 中通常以有结构的对象表示事件。

在无界数据处理中，主要有两类事件概念。

（1）事件时间（Event Time）：事件实际发生的时间。

（2）处理时间（Processing Time）：事件被处理的时间。

并不是所有应用场景都关注事件时间，但其重要性是不言而喻的。例如，在用户行为特征分析、异常检测、基于信贷历史的风控模型等分析中，事件时间起到了决定性的作用。

用户行为特征分析

用户浏览网页或 App 时会产生一系列页面点击与浏览时长数据，这些数据被称为用户行为数据。用户行为数据可以用来判断用户的产品喜好，因此可以用在精准营销、产品功能迭代等环节。

我们可以从这种点击流数据中抽取很多行为指标，如访问频率、平均停留时长、消费行为、信息互动行为、内容发布行为等，从这些指标可以分析出用户的黏性、活跃度，以及产出。

（1）黏性是衡量用户在一段时间内持续访问情况的指标，如访问频率、访问间隔时间。

（2）活跃度指标表征用户访问网页/App 的频次，如平均停留时间、平均访问页面数。

（3）产出指标用来衡量用户的购买价值，如订单数、客单价。

可以对这些指标进一步建模创造商业价值。

异常检测

异常检测（Anomaly Detection）是指找出与大部分对象不同的对象，这种异常对象也被称为离群点。异常检测的应用领域很广，例如可用于数据预处理、病毒木马检测、工业制造产品检测、网络流量检测。

常见的异常检测有以下两类方法。

（1）基于模型的方法：可以通过参数估计的方法创建数据的概率分布模型，以剔除一个不服从该分布的对象。

（2）基于距离的方法：该方法定义对象之间的距离度量，据此检测离群点。

基于信贷历史的风控模型

金融的核心在于控制风险，消费信贷领域也不例外。基于信贷历史的风控模型主要用于反欺诈、判定客户逾期风险、确定用户授信额度。针对消费金融产品的特性，每个公司都有自己的风控模型，这些模型的核心变量是贷款用户的信贷历史数据，包括贷前申请与审核、贷中还款、逾期与催收情况等。除了信贷历史数据，用户个人的其他数据也是模型的输入变量，如个人身份信息（学历、年龄等）、社交信息、电商购物信息等。

在现实世界里，事件时间与处理时间往往并不一致，两者的偏差也因数据源特性、处理引擎及硬件的差别而千差万别，而这种变化给处理引擎的设计带来了不小的挑战。图1-15描述了星球大战系列电影的上映时间。

在这个例子中，事件时间指这是第几部星球大战电影，处理时间指上映时间，如《星球大战1：幽灵的威胁》这部电影的事件时间为1，处理时间为1999年。从图1-15中可以看出，事件时间和处理时间是不一致的。

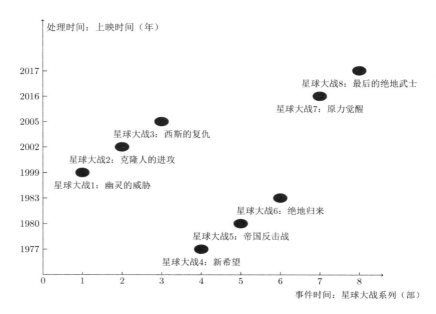

图 1-15 事件时间与处理时间

事件时间与处理时间有所偏差的主要原因如下。

（1）受共享资源局限：如网络阻塞时延、网络分区（参考 CAP 定理）、共享 CPU 等。

（2）软件构架：如分布式系统中的并发竞争、时钟不一致等。

（3）数据自身特性：如 key 的特殊分布、吞吐量的快速涨落、乱序等。

CAP 定理

CAP 定理指出：一个分布式系统最多能具备一致性（Consistency）、可用性（Availability）及分区容错性（Partition tolerance）中的两个特性，而不可能同时具备这三个特性。

（1）一致性：在分布式系统中同一数据的所有备份，在同一时刻，其值是相同的，或者说，所有客户端读取的值是相同的。

（2）可用性：在集群的部分节点出现故障后，集群还能正常响应客户端的读写请求。

（3）分区容错性：分区是对分布式系统通信时限的要求，即如果不能在有限时间内达成数据一致，则系统发生分区。所谓分区容错性，是指即便发生了分区，分布式系统仍然能正确响应客户端的读写请求。

1.3.3 窗口

窗口（window）是将（有界或无界）数据集拆分成一个个有限长度数据区间的机制，即在数据集中增加**临时**处理边界，用于将事件按照时间或其他特征分组分析，其中临时这个定语说明窗口并没有物理地改变数据集。通常有三类窗口，以下举例详细说明。

（1）滚动窗口（Tumbling Window）：将时间拆分成固定长度。如图 1-16 所示，圆点表示事件，属于每个 user（user 1、user 2、user 3）的事件被划到不重叠且等时长（window size）的 5 个窗口（window 1、window 2、window 3、window 4、window 5）中，其中时间（time）可以是事件时间或处理时间。

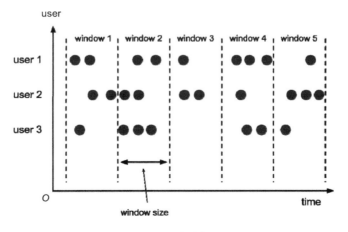

图 1-16　滚动窗口

（2）滑动窗口（Sliding Window）：按照滑动步长（window slide）将时间拆分成固定的长度，当滑动步长小于窗口长度时，相邻窗口间会有重叠。如图 1-17 所示，4 个窗口之间有重叠区域。

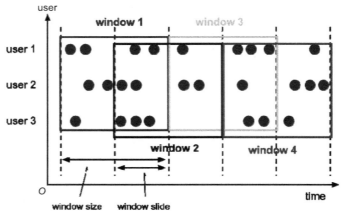

图 1-17　滑动窗口

（3）会话窗口（Session Window）：以活动时间间隔为边界，将一系列连续事件拆分到不同的会话中。会话窗口的长度是动态的。图 1-18 中的 session gap 为窗口的活动时间间隔，user 1 和 user 2 的数据仍被划分为 4 个窗口，但 user 3 的数据被划到了 3 个窗口中。

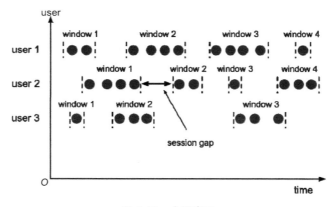

图 1-18　会话窗口

1.3.4 水印

水印（watermark）是嵌入在事件时间轴上用于判断事件时间窗口内所有数据均已到达引擎的一种时间推理工具，是一种既可以在流处理引擎侧嵌入，又可以在消息系统侧嵌入的时间戳。

水印的语义是事件时间小于水印标记时间的事件不会再出现，因此水印是事件的推进器（metric of progress）。

时空穿梭是另一个相关概念。出于调试或审计目的，数据处理程序有时需要将事件时间倒回至过去某个时间点重新开始数据处理任务。

1.3.5 触发器

触发器（trigger）决定在窗口的什么时间点启动应用程序定义的数据处理任务。

水印迟到会拉长窗口生存期，水印早到会导致数据处理结果不准确，触发器就是为解决这两个问题而被引入的。

1.3.6 数据处理模式

1. 有界数据处理

通常，我们使用批处理引擎处理有界数据集（尽管流处理引擎也具备这个功能）。在 Hadoop 的 WordCount 程序中，WordCountMapper 负责遍历数据集的每一行，切分出以空格为间隔的单词，并输出格式为(word,1)的中间处理数据；WordCountReducer 读入 shuffle 后的中间数据集，聚合输出每个单词的出现次数，代码如下：

```
public class WordCountMapper extends Mapper<LongWritable, Text, Text, IntWritable>{
    @Override
    protected void map(LongWritable key, Text value,Context context)
            throws IOException, InterruptedException {
```

```
        String line = value.toString();
        String[] words=line.split(" ");
        for(String word:words){
            context.write(new Text(word), new IntWritable(1));
        }
    }
}
public class WordCountReducer extends Reducer<Text, IntWritable, Text, IntWritable>{
    @Override
    protected void reduce(Text key, Iterable<IntWritable> values,Context context)
            throws IOException, InterruptedException {
        int count = 0;
        for(IntWritable value:values){
            count+=value.get();
        }
        context.write(key, new IntWritable(count));
    }
}
```

这类数据处理模式的特征是将有界数据集处理成规整的形式并输出,可用图 1-19 形象地描述。

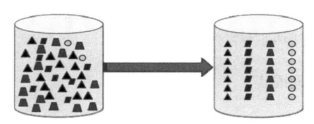

图 1-19 有界数据处理

2. 无界数据批处理

在流处理引擎没有出现之前,业界通常使用窗口机制将无界数据集分割成一系列有界数据块,使用批处理方式分析每个数据块,即微批处理模式,如图 1-20 所示。

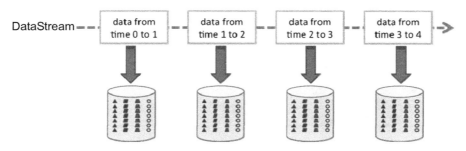

图 1-20　无界数据批处理

3. 无界数据流式处理

基于微批模式的无界数据批处理的原理是，将多个微批处理任务串接起来构建流式数据处理任务，这通常是以牺牲延迟和吞吐为代价的。解决这种缺陷是流处理引擎的发展目标之一，解决方案建立在分析无界数据集的无序和事件时间偏差特性的基础上。

（1）无序意味着引擎需要时间处理机制。

（2）事件时间偏差意味着引擎不能假定在某个时间窗口内能够观察到所有事件。

下面分 4 种情况分析无界数据的流式处理模式。

1）时间无关

流处理引擎能成功应用于需要时间机制的输出处理任务中，也能用于解决与时间无关的数据处理任务，如过滤、数据源连接。

（1）过滤：用于过滤给定规则的事件，如从网站后台日志中过滤出给定源地址的请求。

（2）连接：将两个数据源中相关联的记录连接成一条记录，如图 1-21 所示。

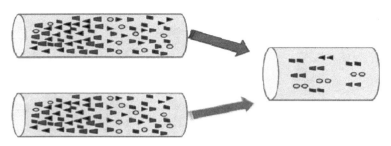

图 1-21　时间无关数据处理

2）近似计算

近似计算是另一种与时间无关的流式数据处理应用，优点是系统开销小。由于算法过于复杂，这类流式数据处理模式应用范围较窄。下面以 Streaming k-means 为例加以说明。

Streaming k-means 算法的基本思想是以空间中 k 个点为中心进行聚类，该算法在流式数据处理应用中的扩展方式如下。

（1）初始以随机位置作为聚类的中心点，因为此时还没有观察到任何数据。

（2）收到新的数据后，调用 Streaming k-means 算法更新中心点。

（3）以更新后的中心点为初始中心点，跳转至步骤 2。

3）根据处理时间开滚动窗口

这种窗口是根据事件被观察的时间设计的，优点有以下 3 个。

- 使用起来简单。
- 窗口边界易于确定。
- 易于提供与事件时间无关的语义。

根据处理时间开滚动窗口的数据处理模式，如图 1-22 所示。

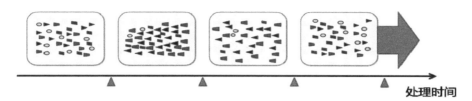

图 1-22　根据处理时间开滚动窗口的数据处理模式

4）根据事件时间开滚动窗口

根据事件时间开滚动窗口的数据处理模式，如图 1-23 所示。

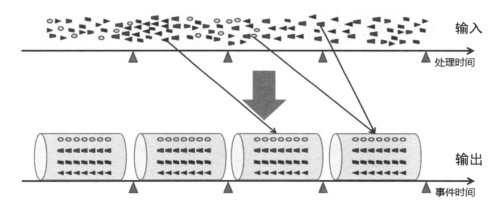

图 1-23　根据事件时间开滚动窗口的数据处理模式

事件时间窗口提供了更灵活的机制，但与处理时间窗口相比，事件时间窗口通常需要更长的生存期，这会给引擎的架构设计带来诸多挑战，例如需要大容量的缓存以持久化状态；窗口边界难以确定，事件迟到导致确定窗口结束点较为困难。

1.3.7　如何理解流式数据架构的内在机制

可以通过以下 4 个问题深入理解流式数据处理的内在机制。

- what：定义数据处理是什么，即对数据进行怎样的转换操作，如 Spark 的算子。
- where：定义转换操作的输入数据取自什么时间区间，窗口是这个问题的答案。
- when：定义转换操作发生在时间轴的什么时间点，水印和触发器是这个问题的答案。
- how：定义如何刻画同一窗口内多次转换操作结果的关系，累加模式是这个问题的答案。

其中，转换操作有时也用聚合代替，转换内涵更丰富，而聚合用于在数据集（如窗口）上计算某个值的场合（如窗口内数据的求和）。

1.4 根据事件时间开滚动窗口

假定某个无界数据集在事件时间区间[12:00, 12:08)内有 10 条记录，每条记录的值都是整数。以事件时间为横轴，以处理时间（观察时间）为纵轴，记录以圆点表示，则所有记录在空间中的位置如图 1-24 所示。为了便于后续引用，这个例子被命名为 Example1.1。

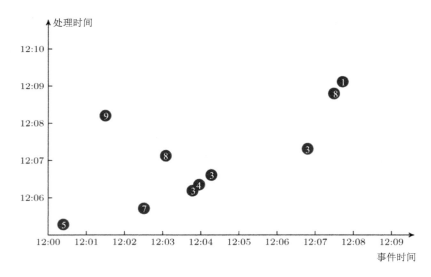

图 1-24　Example1.1 数据分布情况

其中，由于传输通道延迟，处理时间落后于事件时间 5 分钟，即纵坐标轴的零点代表处理时间 12:05。

根据事件时间开滚动窗口，窗口大小为 2 分钟，本节的任务是计算每个窗口内记录值的和。

1.4.1　what：转换 / where：窗口

Example1.1 中的窗口为事件时间窗口，分别为[12:00, 12:02)、[12:02, 12:04)、[12:04, 12:06)和[12:06, 12:08)。转换操作为求和（聚合），聚合结果如图 1-25 所示。

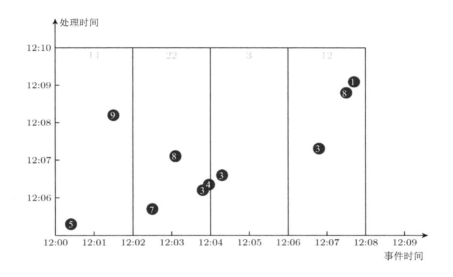

图 1-25　根据事件时间开滚动窗口的聚合结果

聚合操作发生在什么处理时间点上呢？图 1-25 中假定所有聚合操作发生在处理时间点 12:10，水印是解决这个问题的方案。

1.4.2　when：水印

由于事先并不知道横轴的每个窗格里有多少条记录，我们并不能确定何时触发聚合操作，水印则能标记出这个时间点。

水印能够标记出某个事件时间点以前的所有记录均已到达引擎。例如，在窗口[12:00, 12:02)内观察到记录9的同时也观察到代表水印的时间戳12:02，则我们可以在记录9被观察到的时间点上正确地触发聚合操作。

水印可抽象地表示成函数 $f(P) = E$，其中 P 代表处理时间，E 代表事件时间，即我们能够在处理时间点 P 判定事件时间推进到了 E。

图1-26描绘了水印、事件时间和处理时间的关系。

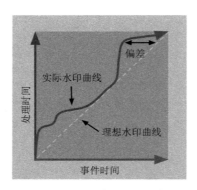

图1-26　水印、事件时间和处理时间的关系

考虑到系统开销，水印是离散的，即只有部分记录后附有水印。为了便于分析，水印通常以连续曲线绘制。此外，作为事件时间的推进器，水印曲线是单调递增的。

有以下两类水印。

（1）完美水印（Perfect Watermark）：完美水印表示早于水印标记事件时间戳的所有记录均已到达，非乱序的无界数据集中最近一条记录的事件时间就是完美水印。

（2）启发式水印（Heuristic Watermark）：启发式水印是尽可能地确定时间戳的一种估计，可能出现某些事件晚于水印到达的情况。在分布式系统中，定义完美水印往往是非常困难的，定义启发式水印的代价则相对较低。

分别嵌入完美水印和启发式水印时窗口的聚合情况，如图1-27所示。

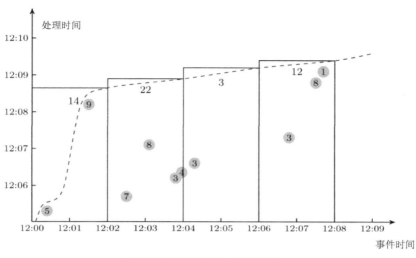

（a）嵌入完美水印时窗口的聚合情况

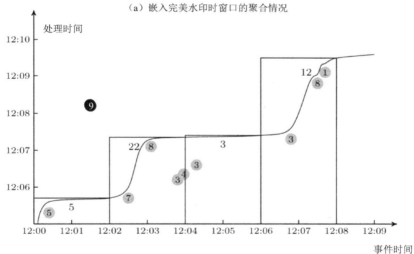

（b）嵌入启发式水印时窗口的聚合情况

图 1-27 分别嵌入完美水印和启发式水印时窗口的聚合情况

可以看出，在嵌入启发式水印时，记录 9 由于水印迟到而没有计入对应事件时间窗口的聚合结果内。

在解决这个问题时，这两类水印都存在缺陷。

（1）水印迟到：在嵌入完美水印时，由于记录 9 在处理时间轴上推进得太慢，事件时间窗口[12:02, 12:04)和[12:04, 12:06)的聚合操作被推迟到处理时间点 12:08 之后，这与低延迟计算的目标相悖；同时，会拉长这两个窗口的生存期，即这两个窗口所占用的资源不能及时释放。

（2）水印早到：在嵌入启发式水印时，事件时间窗口[12:00, 12:02)在处理时间轴上推进得太快，导致记录 9 没能计入本窗口的聚合结果内，这与精准计算的目标相悖，引擎应提供事后更正机制。

1.4.3 when：触发器

作为定义转换操作时间点的另一类方案，触发器解决了水印有缺陷的问题。

（1）在嵌入完美水印时，事件时间窗口[12:02, 12:04)的聚合操作被推迟到处理时间点 12:08 之后，这与低延迟计算的目标相悖。为此，类比将事件时间轴划分为长度为 2 分钟的窗格（pane）而得到事件时间窗口，我们将处理时间轴划分为长度为 1 分钟的窗格，如[12:05, 12:06)、[12:06, 12:07)、[12:07, 12:08)、[12:08, 12:09)、[12:09, 12:10)，然后在每个窗格边界处触发一次聚合计算，这样先后得到**实时**（图 1-28（a）中以"早到"标注）聚合结果 7、14 和 22。由于在得到聚合结果 22 时水印还没有被观察到，这个窗口仍需保留至处理时间点 12:09。

（2）在嵌入完美水印时，事件时间窗口[12:04, 12:06)的聚合操作也被推迟到处理时间点 12:08 之后，按照上述方案需要分别在处理时间窗格[12:05, 12:06)、[12:06, 12:07)、[12:07, 12:08)、[12:08, 12:09)内触发聚合计算。由于这个事件时间窗口内只有处理时间窗格[12:06, 12:07)内有记录，因此在其他窗格内触发聚合计算没有意义，可以通过定义**事件数量触发器**解决这个问题。在本例中可以定义事件数量为 1。

（3）在嵌入启发式水印时，在事件时间窗口[12:02, 12:04)的处理时间窗格[12:07, 12:08)内观察到水印，因此我们不是在处理时间窗格[12:07, 12:08)的边界处触发聚合计算，而是按照水印的推进时间触发聚合计算，这会**按时**（on-time）聚合出结果 22。

（4）在嵌入启发式水印的事件时间窗口[12:00, 12:02)内，水印跟随记录 5 到达，我们会按时得到聚合结果 5。为了得到精准的聚合结果，我们必须延长这个时间窗口的生存期，但是由于并不能确切地获悉还有多少迟到的记录，如何确定这个时间窗口的生存期是个问题。

因此，可以根据水印、处理时间轴窗格和事件数量确定在处理时间轴的什么地方触发聚合计算，如图 1-28 所示。

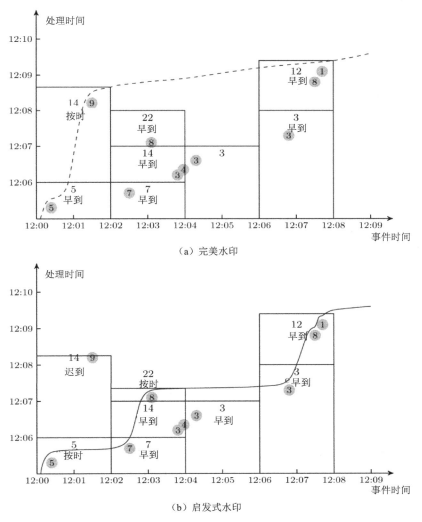

图 1-28　根据事件时间开滚动窗口解决水印有缺陷的问题

截至目前，这个数据处理设计还存在两个需要解决的问题。

（1）在启发式水印早到时，为了确保精准计算，引擎必须延长对应事件时间窗口的生存期，这会加大引擎的内存消耗。

（2）同一个事件时间窗口的多个处理时间窗格会输出多个聚合结果，引擎需要提供定义这些结果之间关系的机制。

可以利用迟到生存期（Allowed Lateness）解决第一个问题，利用累加（Accumulation）模式解决第二个问题。

1.4.4　when：迟到生存期

在嵌入完美水印时，事件不会迟到，窗口能够及时销毁；在启发式水印早到时，为了确保精准计算，引擎必须延长对应事件时间窗口的生存期，所以迟到生存期只会发生在嵌入启发式水印时。

假定迟到生存期为 1 分钟，下面以事件窗口[12:00, 12:02)为例进行分析。

（1）根据水印函数曲线计算出本窗口生存期结束的事件时间点所对应的处理时间点。本窗口生存期结束于事件时间 12:03（12:02+0:01），从水印曲线上找到这个事件时间点对应于处理时间轴上的 12:07~12:08 之间的某个时间点，记为 A。

这里需要再次强调生存期是事件时间，但是决定销毁窗口的时间点是处理时间。在图 1-24 中，我们均匀地标注了事件时间坐标点，如 12:01、12:02 等，但实际上事件时间的推进并不是均匀的，所以我们不能通过处理时间的推进（间隔）推断事件时间的推进，这也是为什么要从水印曲线上找到窗口的处理时间结束点（A）的原因。

（2）如果在处理时间结束点之前观察到事件，则应再次触发聚合计算；在处理时间结束点之后，本窗口被销毁。因此，可以看出**迟到生存期和水印一样都是聚合计算的触发信号**。基于这种定义，记录 9 不会被丢弃。

我们设定迟到生存期为 1 分钟的聚合情况，如图 1-29 所示。

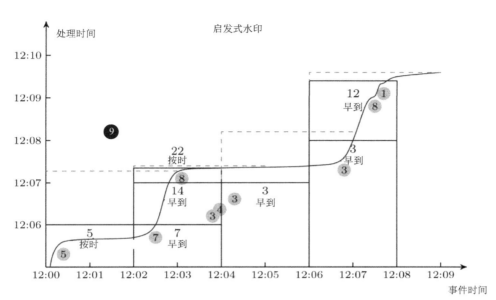

图 1-29　根据事件时间开滚动窗口加入迟到生存期的结果

在图 1-29 中，平行于事件时间轴的虚线标记本窗口的处理时间结束点。

1.4.5　how：累加模式

处理时间轴窗格会多次触发聚合计算，累加模式定义同一个事件时间窗口的多个聚合结果之间的关系。有以下三种累加模式。

- 丢弃（discarding）：启发式水印的事件时间窗口[12:02, 12:04)将产生三个聚合结果，在丢弃的模式下分别为 7、7 和 8，即每个窗格内的聚合结果和其他窗格无关。
- 累加（accumulating）：每个窗格会累加前一个相邻窗格的聚合结果。
- 撤回（retracting）：这种模式是在累加模式的基础上增加一个撤回结果。启发式水印的事件时间窗口[12:02, 12:04)的第二个窗格聚合结果为–7 和 14，–7 代表撤回，14 代表截至当前处理时间总的聚合结果。

这三种模式的聚合情况，分别如图 1-30~图 1-32 所示。

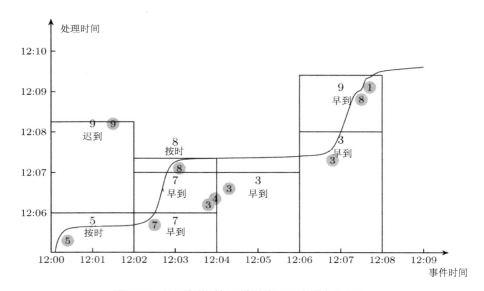

图 1-30　根据事件时间开滚动窗口丢弃模式的结果

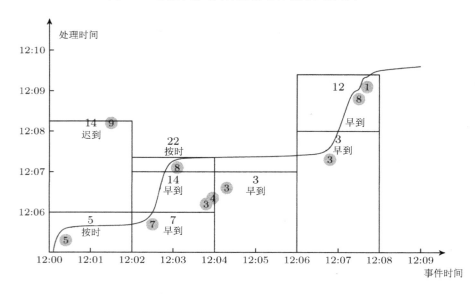

图 1-31　根据事件时间开滚动窗口累加模式的结果

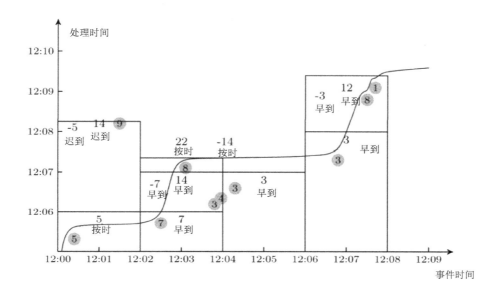

图 1-32　根据事件时间开滚动窗口撤回模式的结果

1.5　一致性

让批处理数据处理程序在由廉价机器组成的集群上可靠地运行不是一件容易的事情，流处理程序则更难。可靠运行的核心问题是如何保证分布式系统有状态计算的一致性。本节将分析在 Flink 架构中容错与一致性的实现方案，即异步屏障快照技术。

1.5.1　有状态计算

在 Example1.1 中，聚合结果与本窗口内的所有记录有关，由于每个记录都是一个独立事件，窗口需要缓存这些独立事件或由这些独立事件产生的中间结果。这类聚合运算被称为有状态计算，而基于单个事件的过滤处理则被称为无状态计算，如图 1-33 所示。

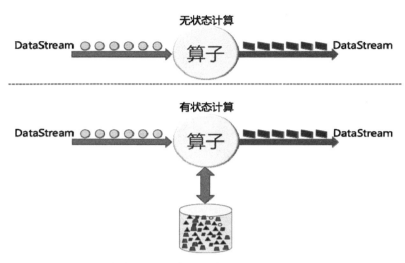

图 1-33　有状态计算与无状态计算

Flink 有以下两类状态。

（1）数据处理应用程序自定义的状态，这类状态由应用程序创建维护。

（2）引擎定义的状态，这类状态由引擎负责管理，如窗口缓存的事件及中间聚合结果。

1.5.2　exactly-once 语义

在分布式系统中，所有数据备份在同一时刻的值是相同的，或者说所有客户端读取的值是相同的，这就是一致性的含义。根据正确性级别的不同，一致性有以下三种形式。

（1）at-most-once：尽可能正确，但不保证一定正确。对应 Example1.1，在系统发生故障恢复后，聚合结果可能会出错。

（2）at-least-once：对应 Example1.1，在系统发生故障恢复后，聚合计算不会漏掉故障恢复之前窗口内的事件，但可能会重复计算某些事件，这通常用于实时性较高但准确性要求不高的场合。例如，Lambda 架构将强实时性的 Storm 和强一致性的 Hadoop 批处理系统融合在一起，Storm 负责实时生成近似结果，Hadoop 负责计算最终精准结果。

（3）exactly-once：对应 Example1.1，在系统发生故障恢复后，聚合结果与假定没有发生故障情况时一致。这种语义加大了高吞吐和低延迟的实现难度，异步屏障快照技术是 Flink 提供这种语义的理论基础。

1.5.3 异步屏障快照

为了更好地理解异步屏障快照（ABS，Asynchronous Barrier Snapshot）理论，我们首先介绍几个相关概念。

（1）检查点（Checkpoint）：关系型数据库并不会立即将提交的事物写回磁盘，而是先写入缓存（Buffer Cache）和重做日志（Redo Log），这种技术能够在保证数据一致性的同时提高数据访问效率。为了提高故障恢复（Crash Recovery）的速度，数据库仅需要回滚某个时间点之后的未写入磁盘的事物，这个时间点就是检查点。

（2）快照（Snapshot）：数据的一个拷贝，有两种实现方式，分别为写时拷贝（COW，Copy On Write）和写重定向（ROW，Redirect On Write），其中 COW 用于读密集型系统，ROW 用于写密集型系统。

（3）消息队列 **pull** 模式：在分布式消息系统中消费者（Consumer）主动连接缓存代理（Broker）获取消息的一种消息消费模式。相应地，在 push 模式中系统将消息主动推送给消费者。

流式数据处理引擎用计算图的形式编译数据处理应用程序，其中计算图用有向无环图（DAG，Directed Acyclic Execution Graph）的形式描述。它有两种表示形式，即逻辑形式和物理部署形式。逻辑形式的计算图由一系列计算节点的单实例组成，而物理部署形式则由计算节点的多个并行实例组成，其中并行实例的含义是在分布式环境中同一计算节点有多个功能相同的物理部署实例，如图 1-34 所示，逻辑形式中的 map() 节点会有两个部署实例 map()[1] 和 map()[2]。

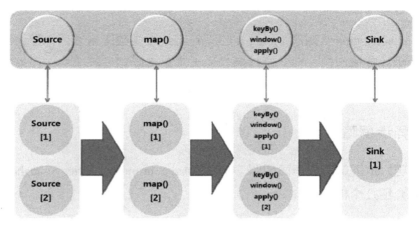

图 1-34　计算图的逻辑形式与物理部署形式

流式数据处理计算图中的节点可分为三类：Source（负责数据输入）、Sink（负责结果输出）和算子（图 1-34 中的 map、keyBy 和 window），它们之间由数据传输通道连接。此外，计算节点的每个部署实例也被称为任务（task）。

以 T 表示计算节点的集合，E 表示边（数据传输通道）的集合，则计算图可表示为 $G = (T, E)$，其抽象形式如图 1-35 所示。

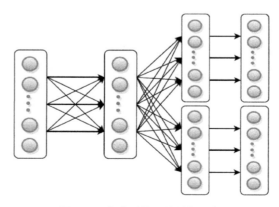

图 1-35　复杂计算图的抽象形式

以 M 表示 E 中传输数据的集合，则对于任意一个计算节点 $t \in T$：

（1）具有输入输出数据集 $I_t, O_t \in E$。

（2）具有状态 s_t。

（3）功能由函数 [这里指用户自定义函数（UDF，User Defined Function）] f_t 定义。节点拉取数据 $r \in M$，由函数 f_t 更新状态至 s'_t，并生成输出数据 $D \in M$，即

$$f_t : s_t, r \quad \mapsto \quad \langle s'_t, D \rangle$$

一个很自然的想法是对计算图 $G = (T, E)$ 在某些时间点上做快照，这样在故障发生后整个数据处理系统可以恢复到某个快照时间点的状态，以保证 exactly-once 语义。

定义快照为

$$G^* = (T^*, E^*)$$

其中，T^* 是所有节点状态的集合，即

$$s_t^* \in T^*, \quad \forall t \in T$$

E^* 是所有传输通道状态的集合，即 $e^* \in E^*$。为了保证 exactly-once 语义，快照 G^* 需要具备以下两个约束条件。

（1）快照必须在有限时间内完成。

（2）快照必须包含所有信息（包括在通道上传输的数据）及这些信息的因果关系，这涉及 T^* 与 E^* 的关系。

为了实现这个看似很自然的想法，先驱者展开了开创性的研究。

（1）同步快照（Global Synchronous Snapshot）。同步快照分为三个步骤：第一步是暂停整个数据处理引擎；第二步是执行快照操作；第三步是继续执行。同步快照包括此刻仍在所有传输通道 $e \in E$ 中的数据和所有节点的状态 s_t，因此快照的容量较大。这种同步机制会严重影响系统的吞吐量，增加引擎运行时的系统开销。

（2）异步快照（Asynchronous Snapshot）。为了规避快照同步造成系统吞吐量降低的缺陷，研究人员提出异步机制，即在引擎执行计算任务的同时执行快照操作，且不需要所有节点和传输通道同时执行快照操作。这种机制并没有解决快照容量大的问题，也没有提升故障恢复效率。

（3）异步屏障快照。这是一种轻量级异步快照，不仅适用于 DAG，而且适用于有环图，本节以 DAG 为例分析快照算法和故障恢复机制。

ABS 的前置条件如下。

（1）传输通道提供阻塞（Block）和非阻塞（Unblock）操作，数据以先进先出（FIFO，First Input First Output）的方式传输。当传输通道处于阻塞状态时，所有数据将被缓存。

（2）计算节点可以阻塞或和与之连接的通道解除阻塞，并能在通道上传输控制消息，还可以在其输出通道上广播消息（Broadcasting Messages）。

（3）控制消息流不参与任何节点函数 f_t 的计算。

快照算法的步骤如下。

（1）引擎定期向 Source 节点插入检查点屏障（Barrier）。在收到作为控制消息的检查点屏障后，Source 节点对自己的状态 s_t 做快照，并在其输出通道上广播此检查点屏障消息。此外，不同的检查点屏障可以通过 id 区分。

（2）当其从任意一个输入通道收到检查点屏障消息时，算子或 Sink 节点阻塞此输入通道，直至本节点从所有输入通道收到检查点屏障。

（3）在其从所有输入通道收到检查点屏障后，算子或 Sink 节点对自己的状态 s_t 做快照，然后对其所有输入通道解除阻塞。

于是，对于同一检查点屏障，ABS 会产生在下面条件时的快照，

$$E^* = \varnothing \tag{1.1}$$

$$G^* = (T^*, E^*)$$

ABS 算法中条件式（1.1）表明快照仅包括节点的状态，不包括仍在传输通道 $e \in E$ 中的数据，因为这会降低快照的容量。同时，ABS 算法很好地兼顾了低延迟和高吞吐。Flink 采用 ABS 算法实现一致性，图 1-36 描述了 Flink 检查点屏障的流转过程。

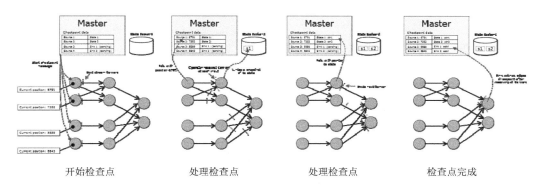

图 1-36　Flink 检查点屏障的流转过程

ABS 是全局的，这样可以通过快照 id 计算出同一时刻计算图 $G^* = (T^*, E^*)$ 的状态，引擎可恢复到这些时间点重启计算任务，进而保证 exactly-once 语义。那么，如何进行故障恢复呢？故障恢复分为以下两种情况。

（1）计算图的拓扑结构不变的情况。引擎从持久化后端中读入上一个可用快照，以此重新初始化所有物理计算任务，即恢复计算图 $G^* = (T^*, E^*)$；然后计算任务继续运行，就如同没有发生故障一样。Flink 窗口的状态不仅包括事件时间推进情况，还包括处理时间推进情况，对于故障恢复这一点异常重要，也从另一个侧面说明引入时间与窗口机制加大了架构流式数据处理引擎的难度。

（2）计算图的拓扑结构发生变化的情况，如图 1-37 所示。

这需要引擎的任务管理器（Job Managers/Task Managers）根据快照重新编排计算任务，这也是弹性（Resilient）计算的要求之一。

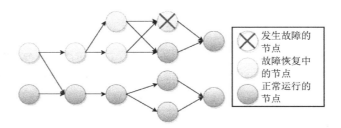

图 1-37　计算图的拓扑结构发生变化的情况

1.5.4　保存点

　　检查点屏障由引擎负责实现，不需要数据处理应用程序编程。保存点（Savepoint）则由应用程序借助检查点底层机制实现一致性的应用层机制，广泛应用于数据处理程序平滑升级中。检查点的目标是轻量，而保存点的目标是实现应用层的一致性功能，例如，Flink 可配置 RocksDB 作为存储后端以实现增量式状态，而保存点则不需要这种优化配置。

　　保存点可用于应用程序平滑升级、引擎升级、A/B 测试等场景。在平滑升级任务中，我们首先保存旧版本程序（如 V1.0）在升级前（如时间点 t_1）的运行状态，然后用保存的状态初始化新版本程序（如 V2.0），这种版本状态管理是由保存点实现的。此外，为了保证平滑升级和升级失败回退，在新版本正常提供服务之前，旧版本仍需继续运行，保存点实现版本状态管理的过程，如图 1-38 所示。

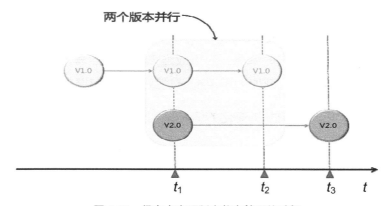

图 1-38　保存点实现版本状态管理的过程

1.6 思考题

（1）如果有两个输入通道的算子 $t \in T$ 从输入通道 $e \in E$ 收到了检查点屏障 b，同时，从通道 $e' \in E$ 收到了属于另一个时刻的检查点屏障 b'，那么这种不同步会不会产生一致性错误呢？

（2）本章分析了根据事件时间开滚动窗口的情况，怎么分析其他窗口机制呢？例如，根据处理时间开滚动窗口、根据事件时间开会话窗口。

（3）故障恢复后，窗口的运行会出现什么情况？在根据事件时间开滚动窗口的例子中，在处理时间 12:07 时刻系统出现故障，并在两分钟后恢复，分析此后的聚合过程。

（4）在嵌入完美水印时，事件不会迟到，窗口能够及时销毁；在嵌入启发式水印早到时，为了确保精准计算，引擎必须延长对应事件时间窗口的生存期，即迟到生存期。既然我们很难生成完美水印，为什么所有窗口都加大迟到生存期，这样在流式数据架构理论中就不需要水印这个概念了吗？

（5）实时机器学习会给流式数据架构的设计带来什么困难？

第 2 章
编程基础

为了让读者快速了解 Flink，2.1 节将介绍 Flink 的特征及其提供的所有数据处理应用 API；2.2 节将介绍如何搭建流处理和批处理 IDEA 开发环境，并介绍如何启动 Flink 自带的 SocketWindowWordCount 流处理程序；2.3 节将详细介绍 Flink 的编程模型，如 API 分层关系、应用程序和 Pipeline 的对应关系，以及应用程序的基本结构；2.4 节将介绍 Flink 运行时，包括运行时结构、任务调度和物理执行计划。

2.1　Flink 概述

从互联网时代的数据爆炸，到即将大规模铺开的 5G 通信支撑的物联网时代的大数据浩海，赋能工具的大规模数据处理技术架构起到了决定性的作用，反过来也推动了技术架构的与时俱进。

作为低延迟、高吞吐、统一流处理和批处理的大数据计算引擎，Flink 正成为实时流式数据处理应用的首选数据处理框架，其主要特征如下：

- 支持高吞吐、低延迟、高性能的流式数据处理，而不是用批处理模拟流处理。
- 支持多种时间窗口，如事件时间窗口、处理时间窗口。

- 支持 exactly-once 语义。
- 具有轻量级容错机制。
- 同时支持批处理和流处理。
- 在 JVM（Java 虚拟机，Java Virtual Machine）层实现内存优化与管理。
- 支持迭代计算。
- 支持程序自动优化。

此外，不仅提供流处理 API、批处理 API，还提供基于这两层 API 的高层数据处理库，包括：

- 机器学习库（FlinkML）。
- 流式关系型 API（Table/SQL）。
- CEP。
- 图分析（Gelly）。

而且，数据处理应用程序可以选择用 Java 语言或 Scala 语言编写，降低了应用程序的编程门槛。

2.2 让轮子转起来

2.2.1 本书约定

（1）本书中的例子以 Scala 语言的编程进行讲解，Flink 的 API 也只讲述 Scala 语言形式。本书中的例子不会过多地运用 Scala 编程技巧，因此读者只需要具备基本的 Scala 语言知识即可。

（2）例子的开发环境为 Java 8（1.8.0_73）、Maven（3.0.4）、SBT（1.2.6）和 Scala（IDEA Scala plugin 2.11.12）。

（3）IDE 选用 IntelliJ IDEA，使用社区版（Community Edition 2018.2.6 x64）。由于存在版本兼容性问题，作者不推荐使用 Eclipse。

（4）开发环境为 Windows 7，Flink 部署环境为 Linux（CentOS）。

（5）Flink 的版本为 1.6.1。

2.2.2 搭建单机版环境

1. 搭建一个单机版的运行环境

（1）下载不带 Hadoop 组件的 Flink 程序包：flink-1.6.1-bin-scala_2.11.tgz。

（2）部署在 Linux 服务器上，然后启动单机版 Flink：

```
# cd flink
# tar xzf flink-*.tgz
# cd flink-1.6.1
# ./bin/start-cluster.sh
```

为了访问方便，设置开发环境机器 hosts 文件，以以下域名映射 Linux 服务器 IP 地址：

```
flink.51deepmind.com
```

（3）启动成功后，在浏览器地址栏中输入以下地址，访问 Flink 的 Web Dashboard：

```
http://flink.51deepmind.com:8081/
```

Web Dashboard 展示当前 Job Manager 和 Task Manager 的状态，如图 2-1 所示。

图 2-1　Flink 的 Web Dashboard

2. 运行 SocketWindowWordCount 程序

（1）启动一个端口号为 9000 的 Socket server：

```
# nc -l 9000
```

（2）运行 SocketWindowWordCount 应用程序：

```
# ./bin/flink run examples/streaming/SocketWindowWordCount.jar --port 9000
```

（3）在 Socket server 端手动输入单词，如果一行有多个单词，就在两个单词之间输入空格。输入及对应的聚合结果如图 2-2 所示。

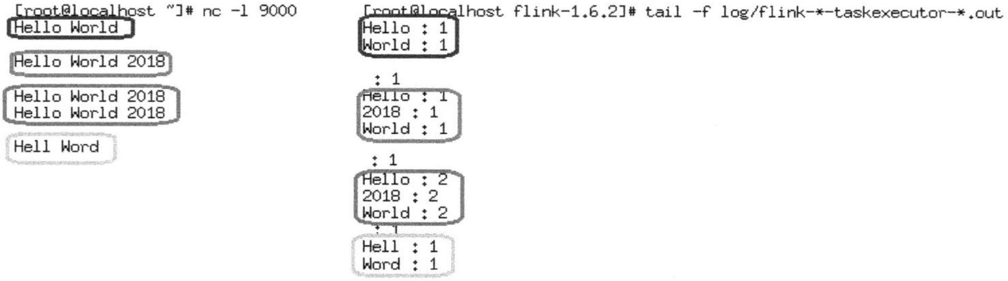

图 2-2 输入及对应的聚合结果

图 2-2 中同一种颜色的输入和输出是对应的，其中 ": 1" 是 Socket server 端换行的聚合结果。

SocketWindowWordCount 应用程序根据处理时间开滚动窗口，每秒计算一次窗口接收单词的次数，代码如下：

```
object SocketWindowWordCount {

    def main(args: Array[String]) : Unit = {

        // the port to connect to
        val port: Int = try {
            ParameterTool.fromArgs(args).getInt("port")
        } catch {
            case e: Exception => {
                System.err.println("No port specified. Please run
```

```
'SocketWindowWordCount --port <port>'")
            return
        }
    }

    // get the execution environment
    val env: StreamExecutionEnvironment =
            StreamExecutionEnvironment.getExecutionEnvironment

    // get input data by connecting to the socket
    val text = env.socketTextStream("localhost", port, '\n')

    // parse the data, group it, window it, and aggregate the counts
    val windowCounts = text
        .flatMap { w => w.split("\\s") }
        .map { w => WordWithCount(w, 1) }
        .keyBy("word")
        .timeWindow(Time.seconds(5), Time.seconds(1))
        .sum("count")

    // print the results with a single thread, rather than in parallel
    windowCounts.print().setParallelism(1)

    env.execute("Socket Window WordCount")
}

// Data type for words with count
case class WordWithCount(word: String, count: Long)
}
```

以上代码从 socket（9000 端口）按行读入字符，切割成单词（w => w.split ("\\s")）后转换成 case 对象（WordWithCount），该对象有两个属性，其中 String 类型属性代表单词本身；Long 类型属性代表单词出现的次数。

其中 timeWindow 为开窗机制，如果应用程序的时间特征为事件时间，则开长度为 5 秒的事件时间窗口，否则开长度为 1 秒的处理时间窗口。Flink 流处理环境（StreamExecutionEnvironment）默认的时间特征为处理时间，因此本例中的开窗机制为长度为 1 秒的处理时间窗口。

2.2.3 配置 IDEA

使用 Maven 从 Flink 官网下载应用程序工程模板。为了避免输入错误，我们设置 Maven 为 Batch 模式，在命令行中设定 groupId、artifactId 和 version，需要注意版本号的值用双引号包裹起来，代码如下：

```
# ./bin/flink run examples/streaming/SocketWindowWordCount.jar --port 9000
mvn archetype:generate \
-DarchetypeGroupId=org.apache.flink \
-DarchetypeArtifactId=flink-quickstart-scala \
-DarchetypeVersion=1.6.1 \
-DgroupId=com.deepmind.flink \
-DartifactId=QuickStart \
-Dversion="1.0-SNAPSHOT" \
-DinteractiveMode=false;  #设置为Batch模式
```

应用程序模板 Maven 的构建过程，如图 2-3 所示。

图 2-3 应用程序模板 Maven 的构建过程

然后，将下载的应用程序工程导入 IDEA。该工程有两个样例程序，分别为批处理应用程序（BatchJob）和流处理应用程序（StreamingJob）。该工程的 pom.xml 文件的主要内容如下：

```xml
<properties>
    <project.build.sourceEncoding>UTF-8</project.build.sourceEncoding>
    <flink.version>1.6.1</flink.version>
    <scala.binary.version>2.11</scala.binary.version>
    <scala.version>2.11.12</scala.version>
</properties>

<dependencies>
    <!-- Apache Flink dependencies -->
    <dependency>
        <groupId>org.apache.flink</groupId>
        <artifactId>flink-scala_${scala.binary.version}</artifactId>
        <version>${flink.version}</version>
        <scope>provided</scope>
    </dependency>
    <dependency>
        <groupId>org.apache.flink</groupId>
        <artifactId>flink-streaming-scala_${scala.binary.version}</artifactId>
        <version>${flink.version}</version>
        <scope>provided</scope>
    </dependency>

    <!-- Scala Library, provided by Flink as well. -->
    <dependency>
        <groupId>org.scala-lang</groupId>
        <artifactId>scala-library</artifactId>
        <version>${scala.version}</version>
        <scope>provided</scope>
    </dependency>

    <!-- These dependencies are excluded from the application JAR by default. -->
    <dependency>
        <groupId>org.slf4j</groupId>
        <artifactId>slf4j-log4j12</artifactId>
        <version>1.7.7</version>
        <scope>runtime</scope>
    </dependency>
```

```xml
<dependency>
    <groupId>log4j</groupId>
    <artifactId>log4j</artifactId>
    <version>1.2.17</version>
    <scope>runtime</scope>
</dependency>
</dependencies>
```

Flink 应用程序模板如图 2-4 所示。

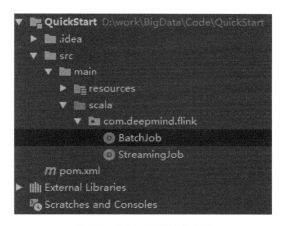

图 2-4　Flink 应用程序模板

此外，为了让开发工具自动检查代码规范，IDEA 开启了 Scala 语言对应的 Checkstyle 功能。

2.3　编程模型

2.3.1　分层组件栈

Flink 的组件分为 4 层，各个模块之间的层次关系如图 2-5 所示。

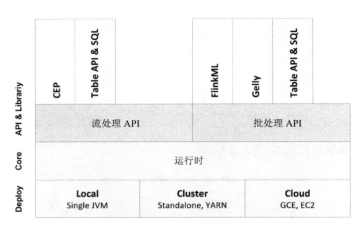

图 2-5　Flink 各个模块之间的层次关系

（1）Deploy 层：Flink 支持多种部署模式，如本地（Local）单机版、Standalone 集群、YARN 集群及云（Cloud）部署模式。

（2）Core 层：本层是 Flink 分布式数据处理引擎的核心实现层，包括计算图 $G=(T,E)$ 的所有底层实现，例如时间与窗口机制、一致性语义、任务管理与调度、物理执行计划。应用程序通常不需要调用本层 API，而是调用流处理 API、批处理 API 或构建在这两层 API 基础之上的 Library API。

（3）API 层：该层包括流处理 API 和批处理 API，Flink 的批处理是建立在流式架构上的，而不是用批处理模拟流处理，这种技术基因决定了 Flink 更适用于流处理的场合。

（4）Library 层：该层是 Flink 的应用框架层，构建在流处理 API 和批处理 API 之上，因此同一应用框架库有两种版本选择，如流式关系型 API（Table/SQL）。此外，本层还包括 CEP、FlinkML 和 Gelly。

2.3.2　流式计算模型

一个典型的流处理应用程序（命名为 Programm 2.1）如下：

```
// 创建运行时
val env: StreamExecutionEnvironment =
StreamExecutionEnvironment.getExecutionEnvironment
// 添加 Source
val lines: DataStream[String] = env
    .addSource(new FlinkKafkaConsumer08[String]("topic", new SimpleStringSchema(),
properties))
// 定义算子转换函数
val events: DataStream[Event] = lines.map { line => parse(line) }
val stats: DataStream[Statistics] = events
        .keyBy("id")
        .timeWindow(Time.seconds(10))
        .apply(new MyWindowAggregationFunction())
// 添加 Sink
stats.addSink(new FlinkKafkaProducer011[String]("localhost:9092","my-topic",
        new SimpleStringSchema))
// 启动程序
env.execute("Kafka Window WordCount")
```

这段程序的逻辑计算图形式如图 2-6 所示。

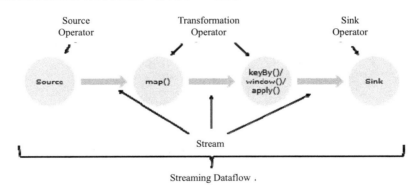

图 2-6　Programm 2.1 的逻辑计算图形式

图 2-6 中 Stream 为传输通道中的数据 $I_t, O_t \in E$，Operator 为计算图的节点 $t \in T$，Streaming Dataflow 为计算图 $G = (T, E)$。

计算图 $G = (T, E)$ 的物理形式由计算节点的多个并行实例组成，其中并行实例的含义是：在分布式环境中，同一计算节点有多个功能相同的物理部署实例，如图 2-7 中逻辑形式中的 map() 节点会有两个部署实例 map()[1] 和 map()[2]。

在并行模式时：

（1）每个 Operator 的实例数为并行度，任意两个 Operator 的并行度之间是独立的。例如，图 2-7 中 Source Operator 的并行度为 2，而 Sink Operator 的并行度为 1；每个 Operator 称为一个任务，Operator 的每个实例称为子任务（subtask），子任务这个概念来自其和 JVM 线程之间的关系。

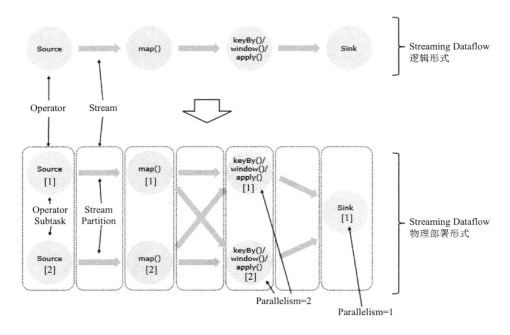

图 2-7　Programm 2.1 的物理计算图形式

（2）Stream 有一个或多个分区（partition）。Stream 有两种模式：

- 直连（One-to-One）模式，即一个实例的输出是另一个实例的输入。在 Programm 2.1 的物理计算图形式中，Source 的 subtask[1]（即 Source[1]）和 map 的 subtask[1]（即 map [1]）直接相连，Source[1] 的输出全部传输给 map [1]，没有被拆分成多个分区。
- 分区（Redistribution）模式，即一个实例的输出被拆分成多个部分传输给多个下级实例。在 Programm 2.1 的物理计算图形式中，map [1] 被拆分成两部分，分别输入给不同的下级实例。

2.3.3 流处理编程

1. DataStream 与 DataSet

Flink 用 DataStream 表示无界数据集，用 DataSet 表示有界数据集，前者用于流处理应用程序，后者用于批处理程序。根据所处理事件数据结构类型的不同，应用程序可以定义不同类型的 DataStream 对象和 DataSet 对象。以下程序定义事件类型为 String 的 DataStream 对象和事件类型为 LabeledVector（带标签的训练样本，每个样本用向量表示）的 DataSet 对象：

```
val lines: DataStream[String] = ...
val input: DataSet[LabeledVector] = ...
```

从操作形式上看，DataStream 和 DataSet 与集合（Collection）有些相似，但是两者有着本质不同：

（1）DataStream 和 DataSet 是不可变的数据集合，因此不可以像操纵集合那样增加或删除 DataStream 和 DataSet 中的元素，也不可以通过诸如下标等方式访问某个元素。这里重申之前定义的概念，事件、元素、数据等都是用于指代流处理或批处理所处理的数据对象的，具体使用哪个称呼依赖语境。

（2）Flink 应用程序通过 Source 创建 DataStream 对象和 DataSet 对象，通过转换操作产生新的 DataStream 对象和 DataSet 对象。

2. 程序结构

Flink 按照数据处理流程编写应用程序，共分为 5 个步骤。

1）获取运行时

运行时分为两类：StreamingExecutionEnvironment 和 ExecutionEnvironment，分别对应流处理和批处理程序：

```
// 流处理运行时
val env: StreamExecutionEnvironment =
StreamExecutionEnvironment.getExecutionEnvironment
// 批处理运行时
val env: ExecutionEnvironment = ExecutionEnvironment.getExecutionEnvironment
```

运行时是应用程序被调度执行时的上下文环境，上述方法根据当前环境自动选择本地或集群运行时环境。以流处理为例，创建方法如下：

（1）通过 createLocalEnvironment 方法创建运行时，基于这种运行时的应用程序会运行在同一个 JVM 进程中，本地调试时通常采用这种运行时。createLocalEnvironment 有三种接口形式：

```
// 有并行度参数设置的 Local 运行时创建方法
 def createLocalEnvironment(parallelism: Int = JavaEnv.getDefaultLocalParallelism):
    StreamExecutionEnvironment = {
  new StreamExecutionEnvironment(JavaEnv.createLocalEnvironment(parallelism))
 }
// 除了并行度，还有带上下文配置的 Local 运行时创建方法
 def createLocalEnvironment(parallelism: Int, configuration: Configuration):
StreamExecutionEnvironment = {
   new StreamExecutionEnvironment(JavaEnv.createLocalEnvironment(parallelism,
configuration))
 }
// 创建 Local 运行时，并启动任务监控 Web
 def createLocalEnvironmentWithWebUI(config: Configuration = null):
StreamExecutionEnvironment = {
    val conf: Configuration = if (config == null) new Configuration() else config
    new StreamExecutionEnvironment(JavaEnv.createLocalEnvironmentWithWebUI(conf))
 }
```

从上面的接口可以看出，通过 createLocalEnvironment 方法创建的运行仍是 StreamingExecutionEnvironment。

（2）通过 createRemoteEnvironment 创建运行时，基于这种运行时的应用程序会被提交到集群中运行，连接集群调试通常用这种运行时。createRemoteEnvironment 有两种接口形式：

```
// 有目标集群 ip、端口和应用程序包的 Remote 运行时创建方法
```

```
def createRemoteEnvironment(host: String, port: Int, jarFiles: String*):
  StreamExecutionEnvironment = {
    new StreamExecutionEnvironment(JavaEnv.createRemoteEnvironment(host, port,
jarFiles: _*))
}
// 除了目标集群ip、端口和应用程序包，还有并行度参数的Remote运行时创建方法
def createRemoteEnvironment(
    host: String,
    port: Int,
    parallelism: Int,
    jarFiles: String*): StreamExecutionEnvironment = {
  val javaEnv = JavaEnv.createRemoteEnvironment(host, port, jarFiles: _*)
  javaEnv.setParallelism(parallelism)
  new StreamExecutionEnvironment(javaEnv)
}
}
```

2）添加外部数据源

可以添加外部数据源，如 Kafka 和文件，也可以由应用创建 DataStream 或 DataSet，后一种方法常用于测试环境。

```
// 通过Kafka创建DataStream
val input: DataStream[String] = env
    .addSource(new FlinkKafkaConsumer08[String]("topic", new SimpleStringSchema(),
properties))

// 通过文件创建DataStream
val input: DataStream[String] = env.readTextFile("file:///path/to/file")

// 通过fromElements创建DataStream
case class Stock(price:Int, volume:Int){
  override def toString: String = price.toString + "_" + volume.toString
}
val input: DataStream[Stock] = env.fromElements(
  "100,1010",
  "120,700"
).map({
  x =>
    val y: Array[String] = x.split(",")
    Stock(y(0).trim().toInt, y(1).trim().toInt)
})
```

3）定义算子转换函数

下面的代码将 input 元素值转换成整型，转换后得到 DataStream[Int]：

```
val mapped: DataStream[Int] = input.map { x => x.toInt }
```

4）定义 Sink

Sink 的功能是将数据处理结果写入外部系统：

```
// 写入外部文件
writeAsText(path: String)
// 打印到控制台
print()
```

除了上述两种常用的 Sink，应用程序还可以将处理结果写入 Kafka：

```
// 写入Kafka
stats.addSink(new FlinkKafkaProducer011[String]("localhost:9092","my-topic",
        new SimpleStringSchema))
```

5）启动程序

调用运行时的 execute() 方法：

```
// 启动程序
env.execute("Kafka Window WordCount")
```

3. 指定键（key）

可以通过 Scala Case 类（或 Java 元组）的位置索引、对象属性名称、key 的选择器（selector）三种方式指定 key，定义如下：

```
// 根据位置指定key, 0代表第一个位置，即Int 类型的属性
val input: DataStream[(Int, String, Long)] = // [...]
val keyed = input.keyBy(0)

// 根据Case 类的属性名称指定key, "word"代表WC对象String 类型的属性
case class WC(word: String, count: Int)
val words: DataStream[WC] = // [...]
val wordCounts = words.keyBy("word").window(/*window specification*/)
```

```
// 通过KeySelector指定key。KeySelector的输入为单个元素，输出为key的位置
// 输出属性word的位置
case class WC(word: String, count: Int)
val words: DataStream[WC] = // [...]
val keyed = words.keyBy( _.word )
```

4．并行度设置

有 4 种设置 Flink 并行度的方式。

（1）通过紧跟在 Operator 之后的 setParallelism 方法设置并行度，这种并行度只影响对应的 Operator：

```
// 设置sum()的并行度为2
val windowCounts = text
    .flatMap { w => w.split("\\s") }
    .map { w => WordWithCount(w, 1) }
    .keyBy("word")
    .timeWindow(Time.seconds(5), Time.seconds(1))
    .sum("count").setParallelism(2)
```

（2）通过运行时设置作业级并行度：

```
env.setParallelism(2)
```

（3）通过客户端设置并行度，这种并行度也是作业级的：

```
try {
    PackagedProgram program = new PackagedProgram(file, args)
    InetSocketAddress jobManagerAddress =
            RemoteExecutor.getInetFromHostport("localhost:6123")
    Configuration config = new Configuration()
    Client client =
            new Client(jobManagerAddress, new Configuration(), program.
getUserCodeClassLoader())
    // 设置作业的并行度为10
    client.run(program, 10, true)
} catch {
    case e: Exception => e.printStackTrace
}
```

（4）通过 Flink 的配置文件设置系统级并行度，这种并行度对集群上的所有

作业都起作用：

```
# 在./conf/flink-conf.yaml 文件中，配置 parallelism.default 值为 10
parallelism.default: 10
```

2.4 运行时

2.4.1 运行时结构

1. Task 线程

线程是程序运行时的最小单元，是进程中的一个实体，是被系统独立调度的基本单位，同属一个进程的所有线程共享进程所拥有的全部资源。

Flink 的每个 Operator 称为一个任务（task），Operator 的每个实例称为子任务，每个任务（包括子任务）在一个 JVM 线程中执行。可以将多个子任务链接（chain）成一个任务，在一个线程中执行，这会降低线程上下文切换产生的开销，减小缓存容量，提高系统吞吐量的同时降低延迟。此外，这种链接机制是可配置的，这增强了数据处理应用程序的灵活性。

> **机制与策略**：理论指导引擎架构实现，但是实现通常考虑得更多。架构通常要权衡复杂度和灵活性，即引擎在简化复杂度的情况下提供什么机制（提供什么能力）、应用程序将会获得更丰富的实现策略（如何使用这些能力）。

在 Programm 2.1 中，Source[1]和 map()[1]、Source[2]和 map()[2]分别链接成一个任务，这种编排是因为 Source 和 map 的两个实例之间采用直连模式，它们之间的数据传输可以通过缓存而不是网络通信，正是这种针对性的优化提升了 Flink 的执行效率。整个应用程序由 5 个线程构成，如图 2-8 所示。

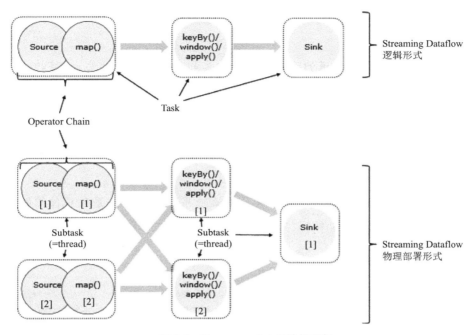

图 2-8　Programm 2.1 的执行线程

从图 2-8 中可见，应用程序由 Operator 的线程组成，也被称为作业（Job）。其中同一作业的一个数据传输通路也被称为一个管道，它用于连接多个命令，将一个命令的执行结果输出给下一个命令。如 Source[1]、map()[1]、keyBy()/window()/apply()[1]和 Sink[1]。

管道这个概念最早出现在 UNIX 操作系统中，下面是 UNIX（类 UNIX 系统）的一个例子：

```
who | tr 'a-z' 'A-Z' > /tmp/who.out
```

上面的命令将当前登录系统用户的信息转换为大写后保存至 /tmp/who.out 文件中，输出结果如下：

```
[root@localhost ~]# who | tr 'a-z' 'A-Z' > /tmp/who.out
[root@localhost ~]# cat /tmp/who.out
ROOT     TTY1         2019-01-24 18:08
```

```
ROOT     PTS/0           2019-01-24 18:08 (192.168.56.1)
[root@localhost ~]#
```

UNIX 系统借用管道的模式将多个独立的命令内聚在一起，以链的方式将它们串接成处理第一个命令（who）输出文本信息的工作流，以降低软件模块（如本例中的命令）之间的耦合。

这是一种重要的软件设计模式，实现了组件之间的"高内聚，低耦合"，降低了模块间协同编程的难度。

这种设计模式在数据处理中有着广泛应用，如何将一系列可伸缩的并行算子编排起来解决目标计算任务，是数据处理引擎的主要任务之一。

2．Manager 进程

Flink 由两类运行时 JVM 进程（process）管理分布式集群的计算资源。

（1）JobManager 进程负责分布式任务管理，如任务调度、检查点、故障恢复等。在高可用（HA，High-Availability）分布式部署时，系统中可以有多个 JobManager，即一个 leader 加多个 standby。JobManager 是 Flink 主从架构中的 master。

（2）TaskManager 进程负责执行任务线程（Programm 2.1 物理部署形式中的 Subtask），以及缓存和传输 stream。TaskManager 是 Flink 主从架构中的 worker。

此外，作为作业的发起者，客户端（client）向 JobManager 提交作业，但客户端并不是 Flink 运行时的一部分，图 2-9 描述了这三者的关系。

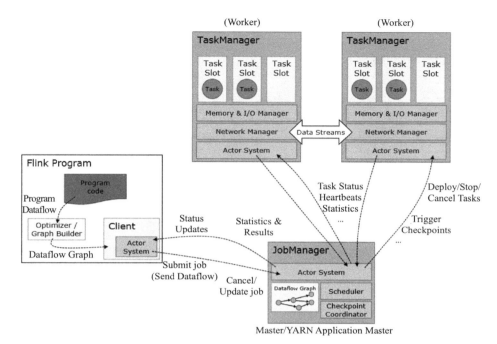

图 2-9　client、JobManager、TaskManager 之间的关系

3．线程共享 Slot

为了控制执行的任务数量，TaskManager 将计算资源划分为多个 Slot，每个 Slot 独享给其分配的计算资源（如内存），这种静态的资源管理方式有利于任务间资源隔离。

TaskManager 可以配置成单 Slot 模式，这样这个 worker 上运行的任务就独占了整个 JVM 进程；同一个 JVM 进程上的多个任务可以共享 TCP 连接、心跳和数据。

Flink 不允许属于不同作业的任务共享同一个 Slot，但允许属于同一个作业的不同任务共享同一个 Slot，因此同一个作业的所有任务可共享同一个 Slot，如图 2-10 所示。

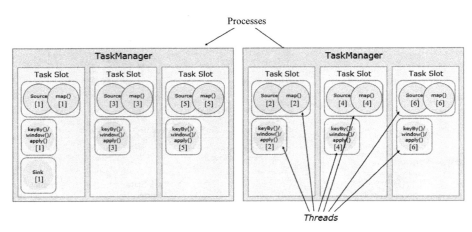

图 2-10　TaskManager 的 Task Slot

2.4.2　任务调度

1．调度策略

以下代码片段对应一个包括 Source、map 和 reduce 的 Pipeline：

```
env.addSource(...).setParallelism(4)
   .map(...).setParallelism(4)
   .reduce(...).setParallelism(3)
```

其中，Source 和 map 的并行度设置为 4（setParallelism(4)），reduce 的并行度设置为 3，Source 和 map 实例间采用直连模式，每个 map 和所有 reduce 均有连接。这个 Pipeline 被调度在两个 TaskManager 上执行，其中每个 TaskManager 有 3 个 Slot。

出于提升执行效率的考虑，Flink 的任务调度系统会并发地执行流处理 Pipeline 中的任务，批处理通常也是如此。应用这种调度策略，本例的调度结果如图 2-11 所示。

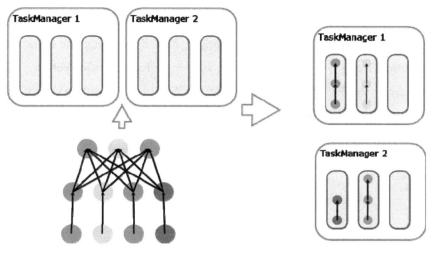

图 2-11　Flink 任务的调度结果

Flink 通过 CoLocationGroup 和 SlotSharingGroup 管理任务的调度约束关系，CoLocationGroup 规定哪些任务必须被调度在同一个 Slot 上运行，而 SlotSharingGroup 则定义哪些任务可以被调度在同一个 Slot 上运行，Flink 的任务调度系统可以根据集群资源使用情况最优化地调度执行作业任务。

2．作业控制

JobManager 将计算图的逻辑形式（JobGraph）编译成物理部署形式（ExecutionGraph）：

（1）JobGraph 由 Operator 和传输通道的数据缓存（Intermediate Data Set）组成。其中，Operator 是计算图中的顶点（JobVertex），并行度控制其实例数量，处理函数（ProcessFunction）定义转换函数 f_t。

（2）ExecutionGraph 由 ExecutionVertex 和 Intermediate Result 的多个分区组成。每个作业的 JobVertex 都对应一个 ExecutionJobVertex，一个 ExecutionJobVertex 对应多个并行 ExecutionVertex 实例；数据缓存也被拆分成多个分区，即 Intermediate Result Partition。

例如，一个 JobGraph 有 4 个顶点，分别记为 JobVertex(A)、JobVertex(B)、JobVertex(C)和 JobVertex(D)；每个顶点的输出都会有 Intermediate Data Set。在对应的 ExecutionGraph 中，每个 JobVertex 对应一个 ExecutionJobVertex，包括每个 JobVertex 的所有并行实例，如 JobVertex(A)对应 ExecutionVertex A(0/2)和 ExecutionVertex A(1/2)。整个作业控制的数据结构如图 2-12 所示。

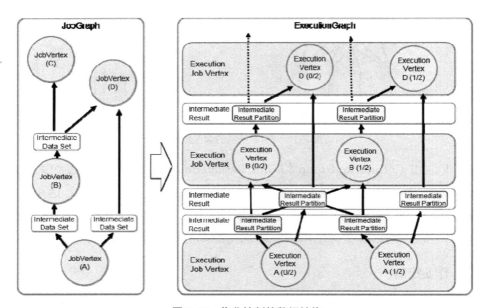

图 2-12　作业控制的数据结构

Flink 通过状态机管理 ExecutionGraph 的作业执行进度。在被创建时，作业的状态为 Created，然后被调度执行，作业的状态流转至 Running，在所有 JobVertex 正常执行完处理任务后，作业结束，即处于 Finished 状态。此外，作业在执行过程中可能会出错，这时状态会流转至 Failing、Restarting、Canceled 或 Failed；作业也可能被 Client 取消，这时状态可能会流转至 Cancelling、Suspended 或 Canceled。作业的状态机转换过程，如图 2-13 所示。

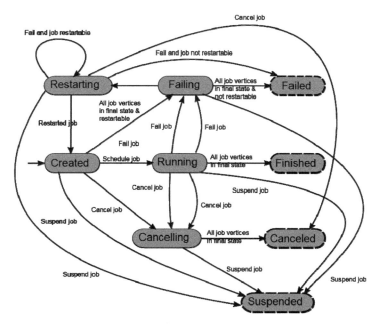

图 2-13 作业的状态机转换过程

2.4.3 物理执行计划

我们可以通过 Web 控制台观察作业的物理执行计划。以 SocketWindowWordCount 为例，单击 Web Flink Dashboard 的 "Add New+" 按钮来添加任务，如图 2-14 所示。

图 2-14 添加任务

选择打包好的 SocketWindowWordCount 程序，并输入参数（--port 9000），单击"Show Plan"按钮，则可获取该程序的物理执行计划，如图 2-15 所示。

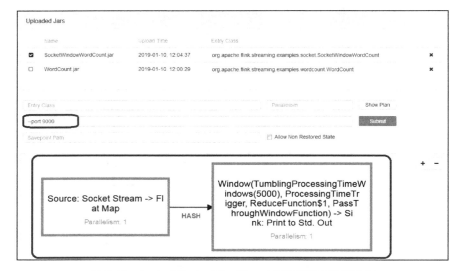

图 2-15　SocketWindowWordCount 程序的物理执行计划

2.5　思考题

（1）Flink 如何隔离多个作业任务？

（2）Flink 如何管理内存？

（3）从 Flink 的分层组件栈图中，我们发现 Table API&SQL 基于 DataStream 和 DataSet 构建流处理和批处理的 API 版本，为什么不在同一个抽象中间层上构建呢？

（4）SocketWindowWordCount 的物理执行计划图中 map 和 window 节点间连线的散列有什么含义？

（5）在一个只包括 map 和 reduce 算子的任务中，Hadoop 和 Flink 的处理机制相同吗？

第 3 章 流处理 API

流处理 API 是流式架构数据处理理论的实现，包括时间、水印、窗口、触发器、迟到生存期、状态与检查点、Source 与 Sink、算子等内容。

3.1 节将概述流处理 Pipeline，并介绍 Source 和 Sink 的常见类型；3.2 节将介绍时间特征设置与水印生成器；3.3 节将介绍算子，包括常见的算子函数、流处理程序分布式部署时的数据分区与资源共享管理；3.4 节将介绍窗口机制，包括窗口分类、窗口函数、触发器、清除器和迟到生存期。窗口是另一类算子，比一般算子具有更丰富的语义，因此将窗口单独成节；3.5 节将介绍连接器，连接器是实现 Source 和 Sink 连接外部系统的关键组件；3.6 节将介绍状态的内部数据结构及其管理方式；3.7 节将介绍如何编程配置检查点。

3.1 流处理 API 概述

流处理程序从 Source 拉取（pull）数据，通过算子 Pipeline 操纵并输入 DataStream，将结果写入 Sink。流处理程序可运行在集群环境和本地 JVM 进程中，也可以嵌入别的应用程序中。

流处理的分析对象是无界数据集，这种数据集是乱序的且存在事件时间偏差。

部分批处理引擎将多个微批处理任务串接起来构建流式数据处理任务，这种架构通常很难兼顾低延迟和高吞吐，为了弥补这种缺陷，我们引入时间与窗口机制，并在此基础上引入水印、触发器与迟到生存期，以解决这类实时数据分析技术所面临的问题。这些技术的应用丰富了算子的语义，本节介绍 Source 与 Sink 的概况。

1）Source 包括消息队列、Socket、文件及使用连接器连接的外部系统，接口定义如下：

（1）文件数据源。

```
// 按行读取整个文件，path 代表文件路径
readTextFile(filePath: String)
// 定义文件格式 inputFormat
readFile (inputFormat: FileInputFormat[T], filePath: String)
```

（2）Socket 数据源。

```
// 连接到主机（host）的 port 端口，以 delimiter 为换行符号
socketTextStream(hostname: String, port: Int, delimiter: Char = '\n', maxRetry: Long = 0)
```

（3）根据集合创建数据源。

```
// 根据相同数据类型的元素创建 DataStream
fromElements (data: T*)
```

（4）外部系统。

```
// 连接到 Kafka
val lines: DataStream[String] = env
    .addSource(new FlinkKafkaConsumer08[String]("topic", new SimpleStringSchema(), properties))
```

2）Sink 包括文件、控制台和外部系统，接口定义如下：

```
// 写入 csv 文件
writeAsCsv(path: String)
// 写入文本文件
writeAsText(path: String)
```

```
// 打印到控制台
print()
// 通过 Socket 写入外部系统
writeToSocket(
    hostname: String,
    port: Integer,
    schema: SerializationSchema[T])
// 通过 SinkFunction 写入外部系统
addSink(sinkFunction: SinkFunction[T])
```

3.2 时间处理

3.2.1 时间

Flink 定义了三类时间：处理时间（Processing Time）取自执行 Operator 的机器的系统时钟；事件时间（Event Time）由数据源产生；进入时间（Ingestion Time）记录被 Source 节点观察时的系统时间，如图 3-1 所示。

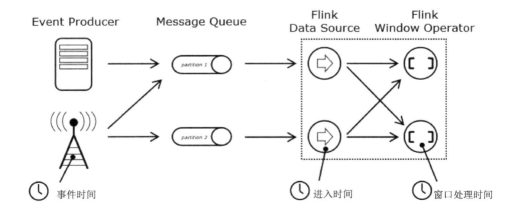

图 3-1　Flink 定义的三类时间

流处理应用程序应显式定义时间特征，用于确定 Source 的时间特性。确定基于系统时间特征的窗口机制，设置方式如下：

```
// 设置事件时间特征
env.setStreamTimeCharacteristic(TimeCharacteristic.EventTime)
// 或者设置为进入时间
// env.setStreamTimeCharacteristic(TimeCharacteristic.IngestionTime)
// 或者设置为处理时间
// env.setStreamTimeCharacteristic(TimeCharacteristic.ProcessingTime)
```

3.2.2 水印

水印是时间戳，由数据源嵌入或由 Flink 应用程序生成。在事件顺序时，水印表明小于其时间戳的事件均已送达且大于其时间戳的事件还没有被观察到。图 3-2 中的数字代表事件时间戳，W(20)代表时间戳为 20 的水印。在水印 W(11)和 W(20)到达之前，没有大于水印时间戳的事件到达。

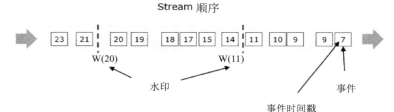

图 3-2　顺序事件的水印

在事件乱序时，大于水印时间戳的事件可能会早于水印到达，例如在水印 W(11)到达之前，事件时间标记为 15 的事件已经到达，如图 3-3 所示。

图 3-3　乱序事件的水印

在分布式执行时，每个实例会将其收到的水印广播到与之相连的下一级节点。图 3-4 中有标记为 A 到 I 的共 9 个事件。

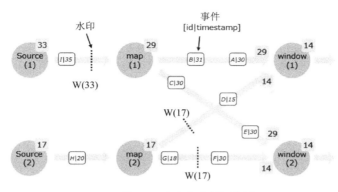

图 3-4 水印在实例间传播

标记的右侧为事件时间，事件时间推进情况会被记录在每一个实例的缓存中，如 map(2) 的当前事件时间戳为 17，这是因为此实例收到了水印 W(17)，并且在收到此水印后将其广播到与 window(1) 和 window(2) 相连的传输通道中。

如果数据源没有正确嵌入水印，则应用程序必须自己生成水印以确保基于事件时间的窗口能够正常工作。为此，DataStream API 提供了自定义水印生成器和内置水印生成器。

3.2.3　周期性水印生成器

周期性水印（Periodic Watermark）根据事件或处理时间周期性地触发水印生成器（Assigner），两个水印时间戳之间并不一定具有固定时间间隔，方法原型如下：

```
def assignTimestampsAndWatermarks(assigner: AssignerWithPeriodicWatermarks[T])
    : DataStream[T]
```

在实现了水印生成器后，应用程序在 Source 节点后的 DataStream 上调用相应生成器，代码如下：

```
val env = StreamExecutionEnvironment.getExecutionEnvironment
env.setStreamTimeCharacteristic(TimeCharacteristic.EventTime)
// 从外部文件中读入 DataStream[Event]
val stream: DataStream[Event] = env.readFile(...)
```

```scala
// 过滤出待处理的数据后调用自定义水印生成器
val withTimestampsAndWatermarks: DataStream[Event] = stream
    .filter( _.severity == WARNING )
    .assignTimestampsAndWatermarks(new MyTimestampsAndWatermarks())

// 转换操作作用域嵌入水印后的 DataStream
withTimestampsAndWatermarks
    .keyBy(...)
    .timeWindow(...)
    .reduce(...)
    .addSink(...)
```

假定最大乱序时间间隔为 3.5s，事件对象定义了获取事件时间属性的方法（getCreationTime），则可通过所有被观察到的事件计算出这些事件的最大时间戳 t，然后发射（emit）一个时间落差为 3.5s 的水印，即此水印的时间戳为 $t - 3.5$。下面是这个例子的实现代码：

```scala
class BoundedOutOfOrdernessGenerator extends AssignerWithPeriodicWatermarks[Event] {
    val maxOutOfOrderness = 3500L // 3.5 seconds
    var currentMaxTimestamp: Long = _
    override def extractTimestamp(element: Event, previousElementTimestamp: Long): Long
= {
        val timestamp = element.getCreationTime
        // 计算最大时间戳
        currentMaxTimestamp = max(timestamp, currentMaxTimestamp)
        timestamp
    }
    override def getCurrentWatermark(): Watermark = {
        // 发射时间落差为 3.5s 的水印
        new Watermark(currentMaxTimestamp - maxOutOfOrderness)
    }
}
```

水印生成器会先调用 extractTimestamp 方法，然后调用 getCurrentWatermark 方法发射水印。这种实现方法是在每个记录后插入水印，相同时间戳的水印不会被发射出去，以确保水印时间戳是严格递增的。

下面是周期性水印的另一个例子：

```
class TimeLagWatermarkGenerator extends AssignerWithPeriodicWatermarks[Event] {
    // 达到 Source 节点的最大延迟时间设定为 5s
    val maxTimeLag = 5000L // 5 seconds
    override def extractTimestamp(element: Event, previousElementTimestamp: Long): Long = {
        element.getCreationTime
    }
    override def getCurrentWatermark(): Watermark = {
        // 水印时间为系统时间减去最大延迟时间
        new Watermark(System.currentTimeMillis() - maxTimeLag)
    }
}
```

以上代码假定事件到达 Source 节点的最大延迟为 5s，在每次事件后发射一个落后于系统时间 5s 的水印。

此外，周期性水印的原型中会带入上一个事件的时间戳：

```
// 上一个事件的时间戳previousElementTimestamp
override def extractTimestamp(element: Event, previousElementTimestamp: Long): Long = {
```

应用程序还可以根据连续两个事件之间的时间关系实现自定义水印。

3.2.4　间歇性水印生成器

间歇性水印（Punctuated Watermark）在观察到事件后，会计算某个条件来决定是否发射水印，方法原型如下：

```
def assignTimestampsAndWatermarks(assigner: AssignerWithPunctuatedWatermarks[T])
    : DataStream[T]
```

假定每个事件均携带事件时间且某些事件会携带代表业务流程结束的标志，如 session 结束标志，则可以在此标志后发射水印，代码如下：

```
class WatermarkOnFlagAssigner extends
         AssignerWithPunctuatedWatermarks[Event] {
```

```
// 抽取事件时间戳
override def extractTimestamp(element: Event, previousElementTimestamp: Long): Long
= {
    element.getSequenceTimestamp
}
// 判断当前事件是否携带结束标志信息
override def checkAndGetNextWatermark(element: Event, previousElementTimestamp)
  : Watermark = {
    Event.isEndOfSequence() ? new Watermark(extractedTimestamp) : null
}
}
```

3.2.5 递增式水印生成器

递增式水印（Assigner with ascending timestamp）生成完美水印，用于顺序的无界数据集。这种水印的含义是：小于其时间戳的事件均已送达且大于其时间戳的事件还没有被观察到。

在使用 Kafka 作为数据源时，每个分区的消息时间通常是递增的，但 Source 节点从多个消息分区并行拉取数据时这种时间特征会被破坏，这时可以在连接器端创建 Kafka 分区水印（Kafka-partition-aware watermark），以确保多分区消息的升序排列，代码如下：

```
// 定义 Kafka 连接器，并创建 Kafka 分区水印
val kafkaSource = new FlinkKafkaConsumer09[MyType]("myTopic", schema, props)
kafkaSource.assignTimestampsAndWatermarks(new AscendingTimestampExtractor[Event] {
    def extractAscendingTimestamp(element: Event): Long = element.eventTimestamp
})

val stream: DataStream[Event] = env.addSource(kafkaSource)
```

Kafka 多消息分区水印，如图 3-5 所示。

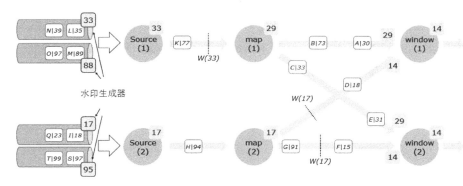

图 3-5 Kafka 多消息分区水印

3.3 算子

简单回顾第 1 章中对计算图的定义。

T 表示计算节点的集合，E 表示边（数据传输通道）的集合，对于计算图 $G = (T, E)$，以 M 表示 E 中传输数据的集合，则对于任意计算节点 $t \in T$：

（1）具有输入输出数据集 $I_t, O_t \in E$，这对应 Flink 运行时 ExecutionGraph 中的传输通道数据缓存（Intermediate Data Set）的多个分区（Intermediate Result Partition）。

（2）具有状态 s_t，在流处理中对应多种状态结构与管理形式，如托管的键控状态、广播状态。

（3）功能由函数 f_t 定义。节点拉取数据 $r \in M$，由函数 f_t 更新状态至 s'_t，并生成输出数据 $D \in M$，即

$$f_t : s_t, r \mapsto \langle s'_t, D \rangle$$

这对应 Flink 运行时 ExecutionGraph 中 ExecutionVertex 定义的功能函数。

因此，DataStream 的算子将会有三个对应的内容：转换函数、数据分区与资源共享管理、状态与容错。

此外，DataStream 的转换结果仍是 DataStream，如 DataStream 在进行 keyby(...) 转换之后得到的是 KeyedStream，而 KeyedStream 则继承 DataStream。因此，应用程序可以在同一个输入 DataStream 上连续使用转换操作，即流畅式 API，编程示例如下所示：

```
input
  .keyBy(...)
  .timeWindow(...)
  .reduce(...)
  .addSink(...)
```

3.3.1 算子函数

1. map

有以下两类 map 操作。

（1）Map：对整个 DataStream 做一一对应的映射，即每一个元素产生一个输出元素。

（2）FlatMap：对整个 DataStream 做一对多的映射，即每一个元素可以产生多个（可以是一个）输出元素。

```
// 将输入元组的两个整型属性相加输入DataStream[Int]
val intPairs: DataStream [(Int, Int)] = // [...]
val intSums = intPairs.map { pair => pair._1 + pair._2 }
// 将输入字符串按空格切分成多个字符串
val textLines: DataStream [String] = // [...]
val words = textLines.flatMap { _.split(" ") }
```

2. KeyBy

对输入 DataStream 分区，即相同 key 的元素分到同一分区，编程示例如下所示：

```
stream.keyBy("someKey")
stream.keyBy(0)
```

3. Reduce

根据多个元素生成一个元素。算子内部会保存上一次 Reduce 的中间结果，然后将当前元素值和上一次 Reduce 的中间结果相加，并用相加结果更新中间结果，编程示例如下所示：

```
keyedStream.reduce { _ + _ }
```

4. 聚合函数

聚合函数根据指定 key 的位置计算，编程示例如下所示：

```
keyedStream.sum(0)
keyedStream.sum("key")
keyedStream.min(0)
keyedStream.min("key")
keyedStream.max(0)
keyedStream.max("key")
keyedStream.minBy(0)
keyedStream.minBy("key")
keyedStream.maxBy(0)
keyedStream.maxBy("key")
```

其中，max(...)返回的是值，而 maxBy(...)返回最大值对应的元素，其他函数与此类似。

5. 窗口函数

窗口函数用于对每个 key 的元素开窗，而 windowAll 则对全体元素开窗，如下代码定义了长度为 5s 的滚动窗口：

```
dataStream.keyBy(0).window(TumblingEventTimeWindows.of(Time.seconds(5)))
dataStream.windowAll(TumblingEventTimeWindows.of(Time.seconds(5)))
```

在窗口内也可以进行 Reduce 和聚合操作，和在 DataStream 上的功能一致，代码如下所示：

```
// 窗口的 Reduce 函数
windowedStream.reduce { _ + _ }
// 窗口的聚合操作
```

```
windowedStream.sum(0)
windowedStream.sum("key")
windowedStream.min(0)
windowedStream.min("key")
windowedStream.max(0)
windowedStream.max("key")
windowedStream.minBy(0)
windowedStream.minBy("key")
windowedStream.maxBy(0)
windowedStream.maxBy("key")
```

6. 连接

1）窗口连接

两个数据源的相同窗口时间区间内元素组合成对，窗口时间区间可以是事件时间或者处理时间，组合对的具体形式也可由 apply 函数定义。这种组合和表内连接相似，下面以滚动窗口说明：

```
val orangeStream: DataStream[Integer] = ...
val greenStream: DataStream[Integer] = ...

// where 用于选择 orangeStream 的 key 的位置，equalTo 用于选择 greenStream 的 key 的位置
orangeStream.join(greenStream)
   .where(elem => /* select key */)
   .equalTo(elem => /* select key */)
   .window(TumblingEventTimeWindows.of(Time.milliseconds(2)))
   .apply { (e1, e2) => e1 + "," + e2 }
```

在图 3-6 中，上下两行数据分别来自不同的数据源。数据源在第 4 个窗口内没有实例，因此没有连接结果输出。

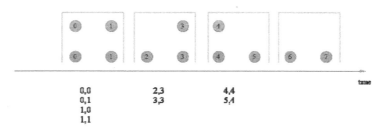

图 3-6　第 4 个窗口内没有实例

2）间隔连接

在事件时间轴上，以被连接数据源（如下列中的 orangeStream）的每一个元素为顶点画锥形，本元素只和被锥形覆盖的另一个数据源的元素组合。其中，锥形的两边分别被定义为下边界（负数）和上边界，即下例中 between 的两个参数：

```
val orangeStream: DataStream[Integer] = ...
val greenStream: DataStream[Integer] = ...

orangeStream
  .keyBy(elem => /* select key */)
  .intervalJoin(greenStream.keyBy(elem => /* select key */))
  .between(Time.milliseconds(-2), Time.milliseconds(1))
  .process(new ProcessJoinFunction[Integer, Integer, String] {
      override def processElement(left: Integer, right: Integer, ctx:
ProcessJoinFunction[Integer, Integer, String]#Context, out: Collector[String]): Unit =
{
      out.collect(left + "," + right);
    }
  });
});
```

间隔连接的情况，如图 3-7 所示。

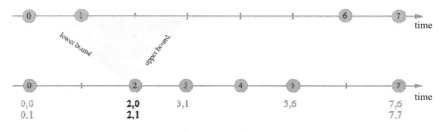

图 3-7　间隔连接的情况

3.3.2　数据分区

分布式系统的通信开销通常都很大，在数据处理应用场景下传输大量数据更是如此。通过合理控制传输通道中的数据分布达到最优的网络通信性能，是实现流式数据处理引擎的一个重要课题，图 3-8 所示为数据分区前后的网络开销对比。

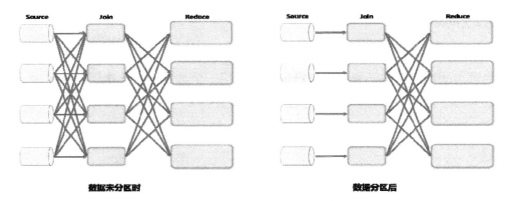

图 3-8　数据分区前后的网络开销对比

在未使用数据分区时，Join 节点的每个并行实例需要聚合来自所有 Source 节点实例的数据，大量数据传输会造成网络过载。使用数据分区后，Source 和 Join 节点实例是一一相连的，可以用同一 Slot 的一个线程运行这两个相连的任务。

Flink 提供 5 种分区 API。

（1）应用程序自定义分区（Custom Partition）：根据指定的 key 位置进行数据分区。

```
dataStream.partitionCustom(partitioner, "someKey")
dataStream.partitionCustom(partitioner, 0)
```

（2）均匀分布分区（Random Partition）：数据会被均匀地分发给下一级节点。

```
dataStream.shuffle()
```

（3）负载均衡分区（Rebalance Partition）：根据轮询调度算法（Round-Robin），将数据均匀地分发给下一级节点。在某些物理拓扑情况下，这是最有效的分区方法，例如 Source 节点和算子节点部署在不同物理设备上。

```
dataStream.rebalance()
```

（4）可伸缩分区（Rescale Partition）：Flink 引擎根据资源使用情况动态调节同一作业的数据分布，根据物理实例部署时的资源共享情况动态调节数据分布，目的是让数据尽可能地在同一 Slot 内流转，以减少网络通信开销。

```
dataStream.rescale()
```

（5）广播分区（Broadcasting Partition）：每一个元素都被广播到所有下一级节点。

```
dataStream.broadcast()
```

3.3.3 资源共享

Flink 将多个任务链接成一个任务在一个线程中执行，在降低线程上下文切换的开销，减小缓存容量，提高系统吞吐量的同时降低延迟。这种机制是可配置的：

（1）创建链。以下代码中后两个 map 函数被链接在一起，而第一个 map 函数则不会被链接。

```
dataStream.map(...).map(...).startNewChain().map(...)
```

（2）关闭作业链接优化，这样任意两个算子实例可不共享线程。

```
dataStream.map(...).disableChaining()
```

（3）Slot 共享组，即在同一个组中所有任务的实例在同一个 Slot 中运行，以隔离非本组实例。

```
dataStream.map(...).slotSharingGroup("name")
```

3.3.4 RichFunction

Flink 定义了一类提供处理函数生命周期及获取函数上下文能力的算子函数，即 RichFunction，其原型如下：

```
public interface RichFunction extends Function {
    // 自定义初始化及销毁处理函数
    void open(Configuration parameters) throws Exception;
    void close() throws Exception;
    // 上下文操作函数
    RuntimeContext getRuntimeContext();
    IterationRuntimeContext getIterationRuntimeContext();
```

```
        void setRuntimeContext(RuntimeContext t);
}
```

其他算子函数实现了对应的版本，如 RichMapFunction、RichReduceFunction、RichJoinFunction 等。

3.3.5 输出带外数据

算子的输出是另一种形式的 DataStream，除此之外，还可以输出自定义的带外数据（Side Output）。带外数据通过名称访问，用以下代码输出带外数据：

```
val input: DataStream[Int] = ...
val outputTag = OutputTag[String]("side-output")

val mainDataStream = input
  .process(new ProcessFunction[Int, Int] {
    override def processElement(
        value: Int,
        ctx: ProcessFunction[Int, Int]#Context,
        out: Collector[Int]): Unit = {
      // emit data to regular output
      out.collect(value)
      // 输出名称为"sideout-"的带外数据
      ctx.output(outputTag, "sideout-" + String.valueOf(value))
    }
  })
```

用以下代码访问带外数据：

```
val outputTag = OutputTag[String]("side-output")
val mainDataStream = ...
// 访问名称为"sideout-"的带外数据
val sideOutputStream: DataStream[String] = mainDataStream.getSideOutput(outputTag)
```

3.4 窗口

窗口是另一类算子，是 DataStream 的逻辑边界，在第一个元素到达后被创建，在生命周期结束后被销毁,但引擎不会确保及时销毁诸如全局窗口(Global Window)这类与时间无关的窗口。除了开窗机制,应用程序还可以定义触发器(Trigger)、迟到生存期、窗口聚合函数(Window Function)及清除器(Evictor)。

窗口分为两大类,即 Keyed Window 和 Non-Keyed Window。在 KeyedStream 上定义 window(...)得到 Keyed Window,在 DataStream 上定义 windowAll(...)得到 Non-Keyed Window。以下是这两类窗口的定义与转换,并不是所有转换应用程序都需要,其中标记为[...]的转换是可选的操作:

```
// Keyed Window 定义与转换
stream
    .keyBy(...)
    .window(...)
   [.trigger(...)]
   [.evictor(...)]
   [.allowedLateness(...)]
   [.sideOutputLateData(...)]
    .reduce/aggregate/…
   [.getSideOutput(...)]

// Non-Keyed Window 定义与转换
stream
    .windowAll(...)
   [.trigger(...)]
   [.evictor(...)]
   [.allowedLateness(...)]
   [.sideOutputLateData(...)]
    .reduce/aggregate/…
   [.getSideOutput(...)]
```

3.4.1 窗口分类

1. 滚动窗口

滚动窗口的时间长度是固定的,且不同时间区间的窗口不会重叠,可根据事件时间和处理时间定义。除了时间长度,还可以设定窗口的时间对齐方式。

如下代码定义的第三个窗口会产生[1:15:00.000，2:15:00.000)、[2:15:00.000，3:15:00.000)之类的时间区间格式的窗口：

```
// 事件时间长度为5s的滚动窗口
input
  .keyBy(<key selector>)
  .window(TumblingEventTimeWindows.of(Time.seconds(5)))
  .<windowed transformation>(<window function>)

// 处理时间长度为5s的滚动窗口
input
  .keyBy(<key selector>)
  .window(TumblingProcessingTimeWindows.of(Time.seconds(5)))
  .<windowed transformation>(<window function>)
// 带起点偏移量的事件时间长度为1h的滚动窗口
input
  .keyBy(<key selector>)
  .window(TumblingEventTimeWindows.of(Time.hours(1), Time.minutes(15)))
  .<windowed transformation>(<window function>)
```

2. 滑动窗口

滑动窗口按照滑动步长将时间拆分成固定长度的窗口，当滑动步长小于窗口长度时，相邻窗口间会重叠，具体代码如下：

```
// 事件时间滑动窗口，窗口长度为10s，滑动步长为5s
input
  .keyBy(<key selector>)
  .window(SlidingEventTimeWindows.of(Time.seconds(10), Time.seconds(5)))
  .<windowed transformation>(<window function>)

// 处理时间滑动窗口，窗口长度为10s，滑动步长为5s
input
  .keyBy(<key selector>)
  .window(SlidingProcessingTimeWindows.of(Time.seconds(10), Time.seconds(5)))
  .<windowed transformation>(<window function>)

// 带起点偏移量的处理时间滑动窗口，窗口长度为12h，滑动步长为1h，对齐偏移量为8h
input
  .keyBy(<key selector>)
```

```
      .window(SlidingProcessingTimeWindows.of(Time.hours(12), Time.hours(1),
Time.hours(8)))
      .<windowed transformation>(<window function>)
```

3. 会话窗口

根据相邻元素之间的时间间隔确定会话窗口的边界,其分为固定时间间隔（Static Gap）和动态时间间隔（Dynamic Gap）两种类型,其中动态时间间隔由应用程序编程实现：

```
// 固定时间间隔为 10s 的事件时间会话窗口
input
    .keyBy(<key selector>)
    .window(EventTimeSessionWindows.withGap(Time.minutes(10)))
    .<windowed transformation>(<window function>)

// 动态时间间隔的事件时间会话窗口
input
    .keyBy(<key selector>)
    .window(EventTimeSessionWindows.withDynamicGap(new
      SessionWindowTimeGapExtractor[Event] {
        override def extract(element: Event): Long = {
          // 根据事件特征确定会话窗口间隔
          ...
        }
    }))
    .<windowed transformation>(<window function>)

// 固定时间间隔为 10s 的处理时间会话窗口
input
    .keyBy(<key selector>)
    .window(ProcessingTimeSessionWindows.withGap(Time.minutes(10)))
    .<windowed transformation>(<window function>)

// 动态时间间隔的处理时间会话窗口
input
    .keyBy(<key selector>)
    .window(DynamicProcessingTimeSessionWindows.withDynamicGap(new
      SessionWindowTimeGapExtractor[Event] {
        override def extract(element: Event): Long = {
          // 根据事件特征确定会话窗口间隔
```

```
    ...
  }
}))
.<windowed transformation>(<window function>)
```

4. 全局窗口

全局窗口将相同 key 的所有元素聚在一起，但是这种窗口没有起点也没有终点，因此必须自定义触发器：

```
input
  .keyBy(<key selector>)
  .window(GlobalWindows.create())
  .<windowed transformation>(<window function>)
```

3.4.2 窗口函数

1. 增量式计算

拆分窗口的主要目的是将指定时间区间内的所有元素当成一个有界数据集，以分析这个数据集的整体特征，如窗口内元素计数、求和。因此，窗口收集到本窗口内所有元素后再去调用窗口函数是一个显而易见的方法，但从数据处理性能上分析，这种方法的计算效率不高。以求和为例，假定某个窗口内有 100 个元素，这些元素间隔为 1s，那么前 99s 窗口一直在收集这些元素。数据处理任务中聚合计算通常需要较多的 CPU，这等价于在前 99s CPU 一直在空转，因此将计算任务均衡地分摊到前 99s 可以减轻第 100s 的计算压力，这被定义为增量式计算。增量式计算增加了数据处理引擎架构设计的复杂度，但换取了计算性能的提升。

2. Reduce 函数

Reduce 函数将每个窗口内的数据作为输入，输出一个计算结果，以下代码计算窗口内所有元素第二个属性的和：

```
val input: DataStream[(String, Long)] = ...

input
  .keyBy(<key selector>)
```

```
.window(<window assigner>)
.reduce { (v1, v2) => (v1._1, v1._2 + v2._2) }
```

3. 聚合函数

聚合函数（AggregateFunction）的 Java 语言原型如下：

```
public interface AggregateFunction<IN, ACC, OUT> extends Function, Serializable {
    ACC createAccumulator();
    ACC add(IN value, ACC accumulator);
    OUT getResult(ACC accumulator);
    ACC merge(ACC a, ACC b);
}
```

聚合函数有三个泛型参数，分别为窗口内数据类型（IN）、累加器类型（ACC）、聚合结果类型（OUT），其中累加器存放临时聚合结果。聚合函数调用 createAccumulator() 初始化累加器，调用 add() 实现增量计算，调用 getResult() 返回聚合结果。在窗口合并时，聚合函数调用 merge() 合并两个窗口的聚合结果。以下是聚合函数的例子：

```
class AverageAggregate extends AggregateFunction[(String, Long), (Long, Long), Double]
{
  // 初始化累加器
  override def createAccumulator() = (0L, 0L)
  // 累加器的第一个参数为元素累加和，累加器的第二个参数为元素数量
  override def add(value: (String, Long), accumulator: (Long, Long)) =
    (accumulator._1 + value._2, accumulator._2 + 1L)
  // 输出平均值
  override def getResult(accumulator: (Long, Long)) = accumulator._1 / accumulator._2
  override def merge(a: (Long, Long), b: (Long, Long)) =
    (a._1 + b._1, a._2 + b._2)
}
val input: DataStream[(String, Long)] = ...
input
   .keyBy(<key selector>)
   .window(<window assigner>)
   .aggregate(new AverageAggregate)
```

4. 处理函数

处理函数（ProcessWindowFunction）可通过迭代器访问窗口内所有的元素，

并可访问窗口上下文，其转换结果是多值的（Collector<OUT>），其原型如下：

```
public abstract class ProcessWindowFunction<IN, OUT, KEY, W extends Window>
                extends AbstractRichFunction {
    public abstract void process(KEY key,
                                 Context context,
                                 Iterable<IN> elements,
                                 Collector<OUT> out) throws Exception;

    public void clear(Context context) throws Exception {}

    public abstract class Context implements java.io.Serializable {
        public abstract W window();
        public abstract long currentProcessingTime();
        public abstract long currentWatermark();
        public abstract KeyedStateStore windowState();
        public abstract KeyedStateStore globalState();
        public abstract <X> void output(OutputTag<X> outputTag, X value);
    }
}
```

窗口上下文包括开窗机制、处理时间、水印、状态等。

下例计算了每一个窗口内元素的数量，并将数量信息发射到输出流中：

```
val input: DataStream[(String, Long)] = ...

input
  .keyBy(_._1)
  .timeWindow(Time.minutes(5))
  .process(new MyProcessWindowFunction())

class MyProcessWindowFunction extends ProcessWindowFunction[(String, Long), String,
String, TimeWindow] {
  def process(key: String, context: Context, input: Iterable[(String, Long)], out:
Collector[String]): () = {
    var count = 0L
    for (in <- input) {
      count = count + 1
    }
    out.collect(s"Window ${context.window} count: $count")
  }
```

}
```

### 5．带窗口函数的聚合函数

WindowedStream 类重载了 aggregate 方法，其中一种重载方法带有 ReduceFunction 和 ProcessWindowFunction 参数，原型如下：

```
public <ACC, V, R> SingleOutputStreamOperator<R> aggregate(
        AggregateFunction<T, ACC, V> aggFunction,
        WindowFunction<V, R, K, W> windowFunction) {
  ...
}
```

在下例中，聚合函数增量式计算均值，ProcessWindowFunction 实时输出均值结果：

```
val input: DataStream[(String, Long)] = ...
input
  .keyBy(<key selector>)
  .timeWindow(<duration>)
  .aggregate(new AverageAggregate(), new MyProcessWindowFunction())

class AverageAggregate extends AggregateFunction[(String, Long), (Long, Long), Double]
{
  // 初始化累加器
  override def createAccumulator() = (0L, 0L)
  // 累加器的第一个参数为元素类累加和，累加器的第二个参数为元素数量
  override def add(value: (String, Long), accumulator: (Long, Long)) =
    (accumulator._1 + value._2, accumulator._2 + 1L)
  // 输出平均值
  override def getResult(accumulator: (Long, Long)) = accumulator._1 / accumulator._2
  override def merge(a: (Long, Long), b: (Long, Long)) =
    (a._1 + b._1, a._2 + b._2)
}

class MyProcessWindowFunction extends ProcessWindowFunction[Double, (String, Double),
String, TimeWindow] {
  def process(key: String, context: Context, averages: Iterable[Double],
          out: Collector[(String, Double)]): () = {
    // 输出当前均值
    val average = averages.iterator.next()
```
```

```
    out.collect((key, average))
  }
}
```

3.4.3 触发器

触发器原型中包括 4 类触发机制，基于事件驱动。

（1）onElement：窗口每收到一个元素调用一次该方法，返回结果决定是否触发算子函数。

（2）onProcessingTime：根据注册的处理时间定时器触发，定时时间由参数 time（long time）设定。

（3）onEventTime：根据注册的事件时间定时器触发，定时时间由参数 time 设定。

（4）onMerge：两个窗口合并时触发。

此外，触发器还提供资源清除接口 clear()，原型如下：

```
public abstract class Trigger<T, W extends Window> implements Serializable {
    public abstract TriggerResult onElement(T element, long timestamp,
                                W window, TriggerContext ctx) throws Exception;
    public abstract TriggerResult onProcessingTime(long time,
                                W window, TriggerContext ctx) throws Exception;
    public abstract TriggerResult onEventTime(long time,
                                W window, TriggerContext ctx) throws Exception;
    public boolean canMerge() {
        return false;
    }
    public void onMerge(W window, OnMergeContext ctx) throws Exception {
        throw new UnsupportedOperationException("This trigger does not support merging.");
    }
    public abstract void clear(W window, TriggerContext ctx) throws Exception;
    public interface TriggerContext {
        long getCurrentProcessingTime();
        MetricGroup getMetricGroup();
```

```
        long getCurrentWatermark();
        void registerProcessingTimeTimer(long time);
        void registerEventTimeTimer(long time);
        void deleteProcessingTimeTimer(long time);
        void deleteEventTimeTimer(long time);
        <S extends State> S getPartitionedState(StateDescriptor<S, ?>
stateDescriptor);
    }
    public interface OnMergeContext extends TriggerContext {
        <S extends MergingState<?, ?>> void mergePartitionedState(
                    StateDescriptor<S, ?> stateDescriptor);
    }
}
```

前三类触发机制的结果（TriggerResult）分为以下 4 种情况。

（1）忽略（CONTINUE）。

（2）触发（FIRE）。

（3）清除（PURGE）：清空窗口内所有元素，窗口被销毁。

（4）触发并清除（FIRE_AND_PURGE）：触发窗口函数，并在函数执行结束后清空窗口内所有元素，窗口被销毁。

Flink 提供几类内置触发器。

- EventTimeTrigger：根据事件时间轴上的水印触发。
- ProcessingTimeTrigger：根据处理时间触发。
- CountTrigger：根据窗口内元素的数量触发。
- ContinuousEventTimeTrigger：将事件时间轴分成等间隔的窗格，在每一个窗格内判断水印来决定是否触发。
- ContinuousProcessingTimeTrigger：将处理时间轴分成等间隔的窗格，在每一个窗格内触发一次，但是需要根据相关条件判断是否调用窗口函数。
- DeltaTrigger：根据某种特征是否超过指定的阈值决定是否触发。
- PurgingTrigger：将其他触发器转化成清除触发器，即销毁窗口。

3.4.4 清除器

清除器（Evictor）在触发器触发后，窗口函数执行前或执行后清除窗口内元素，相应地，有以下两个方法。

第一个方法是在触发器被触发后，窗口函数执行前清除窗口内元素：

```
void evictBefore(Iterable<TimestampedValue<T>> elements, int size, W window,
EvictorContext evictorContext);
```

第二个方法是在触发器被触发后，窗口函数执行后清除窗口内元素：

```
void evictAfter(Iterable<TimestampedValue<T>> elements, int size, W window,
EvictorContext evictorContext);
```

在清除器中，可同时使用这两种方法。

Flink 提供三种内置清除器。

（1）CountEvictor：保持窗口内元素数量为预定值。

（2）DeltaEvictor：根据元素之间的关系，清除超过指定阈值的元素。

（3）TimeEvictor：根据窗口内元素的时间戳决定清除哪些元素。

3.4.5 迟到生存期

Flink 默认的迟到生存期为 0，即事件时间窗口在水印到来后结束，无须考虑事件迟到的情况。

以下代码设置窗口的迟到生存期为 10s：

```
val input: DataStream[T] = ...

input
    .keyBy(...)
    .window(...)
    .allowedLateness(Time.seconds(10))
    ...
```

3.5 连接器

Source 和 Sink 节点连接外部数据源的组件称为连接器（Connector），其中内置连接器的实现代码集成在 Flink 源码中，但是这些代码并没有被编译进 Flink 二进制程序包（Binary Distribution）中。内置连接器如下，其中括号里描述了对应连接器是支持 Source 还是 Sink。

- Apache Kafka（Source/Sink）
- Apache Cassandra（Sink）
- Amazon Kinesis Streams（Source/Sink）
- Elasticsearch（Sink）
- Hadoop FileSystem（Sink）
- RabbitMQ（Source/Sink）
- Apache NiFi（Source/Sink）
- Twitter Streaming API（Source）

此外，并不是所有连接器都支持 exactly-once 语义，尤其是 Sink 节点的连接器。Source 一致性保障情况，如表 3-1 所示。Sink 一致性保障情况，如表 3-2 所示。

表 3-1　Source 一致性保障情况

Source	一致性保障
Apache Kafka	exactly once
AWS Kinesis Streams	exactly once
RabbitMQ	at most once (v 0.10) / exactly once (v 1.0)
Twitter Streaming API	at most once
Collections	exactly once
Files	exactly once
Sockets	at most once

表 3-2　Sink 一致性保障情况

Sink	一致性保障
HDFS Rolling Sink	exactly once
Elasticsearch	at least once
Kafka Producer	at least once
Cassandra Sink	at least once / exactly once
AWS Kinesis Streams	at least once
File Sink	at least once
Socket Sink	at least once
Standard Output	at least once
Redis Sink	at least once

3.5.1　HDFS 连接器

HDFS 连接器是以库的形式提供的，在 Maven 工程中需要加入以下依赖：

```xml
<dependency>
  <groupId>org.apache.flink</groupId>
  <artifactId>flink-connector-filesystem_2.11</artifactId>
  <version>1.6.1</version>
</dependency>
```

为了存储无界数据集，分布式文件系统按照一定规则将 DataStream 拆分成多个分区，每个分区对应一个文件。为了避免数据无限增长，分布式文件系统定义相应的回滚（rolling）策略，代码如下：

```scala
val input: DataStream[Tuple2[IntWritable, Text]] = ...

// 定义文件存储路径
val sink = new BucketingSink[String]("/base/path")
// 按照时间格式命名分区文件
sink.setBucketer(new
          DateTimeBucketer[String]("yyyy-MM-dd--HHmm",
ZoneId.of("Asia/Shanghai")))
// 设置单个文件为 400MB，回滚时长为 20mins
```

```
sink.setWriter(new SequenceFileWriter[IntWritable, Text]())
sink.setBatchSize(1024 * 1024 * 400)
sink.setBatchRolloverInterval(20 * 60 * 1000)

input.addSink(sink)
```

3.5.2 Kafka

Kafka 是一类高吞吐消息系统，由 Scala 和 Java 语言编写，用于大型应用系统组件之间解耦，其架构如图 3-9 所示。

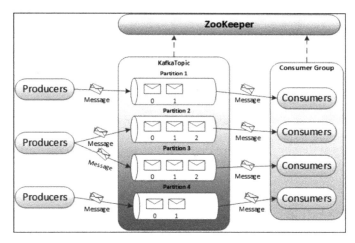

图 3-9 Kafka 的架构

架构说明如下：

（1）Producers 生产消息，Consumers 通过 Broker 订阅消息。

（2）消息通过 Topic 分类，属于同一个 Topic 的消息可分为多个分区。

（3）Consumers 通过位置偏移（offset）消费消息。

（4）ZooKeeper 是集群的分布式协同组件。

Kafka 连接器是以库的形式提供的，在 Maven 工程中需要加入以下依赖：

```xml
<dependency>
 <groupId>org.apache.flink</groupId>
 <artifactId>flink-connector-kafka-0.8_2.11</artifactId>
 <version>1.6.1</version>
</dependency>
```

Kafka 的 Source 连接器编程分为 5 个部分。

1. 创建连接器

创建 Kafka 连接器需要定义 Consumer 属性、Topic 和反序列化器，代码如下：

```scala
val properties = new Properties()
// 设置Broker地址，以逗号分割多个地址
properties.setProperty("bootstrap.servers", "localhost:9092")
// 设置ZooKeeper集群地址，以逗号分割多个地址，kafka 0.8版本独有
properties.setProperty("zookeeper.connect", "localhost:2181")
// 设置Consumer组
properties.setProperty("group.id", "test")
// 创建连接器，消费主题为"topic"，反序列化器为SimpleStringSchema，消息格式为String
val myConsumer = new FlinkKafkaConsumer08[String]
                ("topic", new SimpleStringSchema(), properties)
stream = env
  .addSource(myConsumer)
```

2. 创建反序列化器

消息是有结构的，Flink 应用程序需要定义反序列化器，将消息转换成对象。自定义反序列化器继承抽象反序列化器 AbstractDeserializationSchema，主要接口为 deserialize，用于将字节流反序列化成对象，原型定义如下：

```
T deserialize(byte[] message) throws IOException
```

除了可自己实现反序列化器，还可以使用 avro 反序列化器，在 Maven 工程中需要加入以下依赖：

```xml
<dependency>
 <groupId>org.apache.flink</groupId>
 <artifactId>flink-avro</artifactId>
 <version>1.7-SNAPSHOT</version>
</dependency>
```

3. 设置消息起始位置偏移

设置消息起始位置偏移有以下 4 种方式。

（1）从 Consumer 组上一次提交的位置后开始，以下代码设置了相对于上一次提交的偏移位置。

```
myConsumer.setStartFromGroupOffsets()
```

（2）从最早或最晚的记录开始。

```
myConsumer.setStartFromEarliest()
myConsumer.setStartFromLatest()
```

（3）从某个固定的时间点开始。

```
myConsumer.setStartFromTimestamp(23L)
```

（4）设定每个分区的起始位置。

```
val specificStartOffsets = new java.util.HashMap[KafkaTopicPartition, java.lang.Long]()
// 设置主题"myTopic"的第一个分区起始点为 23L（时间戳，单位为ms）
specificStartOffsets.put(new KafkaTopicPartition("myTopic", 0), 23L)
specificStartOffsets.put(new KafkaTopicPartition("myTopic", 1), 31L)
specificStartOffsets.put(new KafkaTopicPartition("myTopic", 2), 43L)
myConsumer.setStartFromSpecificOffsets(specificStartOffsets)
```

4. 设定检查点周期

Flink 周期性地向 Source 节点插入检查点屏障，如配置检查点周期为 5 秒，代码如下：

```
env.enableCheckpointing(5000)
```

5. 设置位置偏移提交方式

（1）在 Flink 启用检查点机制时，Consumer 在检查点完成后可以将位置偏移提交到 ZooKeeper，但应用程序可以关闭这种功能：

```
// 关闭自动提交功能
myConsumer.setCommitOffsetsOnCheckpoints(false)
```

（2）在没有启用检查点机制时，可以启动自动提交功能：

```
properties.setProperty("enable.auto.commit", "true")
properties.setProperty("auto.commit.interval.ms", "100")
```

3.5.3 异步 I/O

在从外部 Source 拉取数据或将结果写入外部存储时，访问效率和网络通信延迟是流式数据处理引擎所面临的主要构架设计问题之一。使用异步 I/O 意味着多个任务可以并发地访问外部存储，这往往比提高并行度更高效，因为异步 I/O 可以获得硬件层的支持，如 DMA。同步 I/O 和异步 I/O 的对比，如图 3-10 所示。

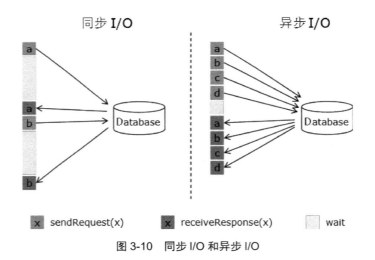

图 3-10 同步 I/O 和异步 I/O

通常，异步 I/O 会带来乱序问题。为此，异步 I/O 客户端需要复杂的处理逻辑以保证请求结果返回的顺序。

Flink 异步 I/O 架构的核心组件为用于实现回调的请求收集器（AsyncCollector）和收集器缓冲区（AsyncCollectorBuffer），应用程序通过 Rich Function 接口调用异步 I/O。异步 I/O 架构，如图 3-11 所示。

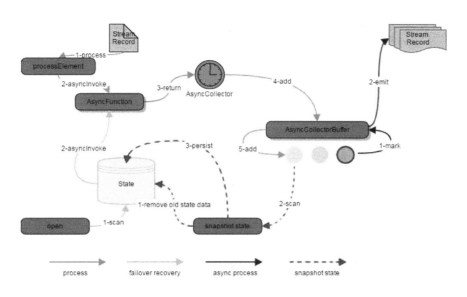

图 3-11　异步 I/O 架构

以下是使用数据库异步 I/O 功能的例子:

```
class AsyncDatabaseRequest extends AsyncFunction[String, (String, String)] {
    // 创建数据库客户端
lazy val client: DatabaseClient = new DatabaseClient(host, post, credentials)
// 创建执行器
implicit lazy val executor: ExecutionContext =
 ExecutionContext.fromExecutor(Executors.directExecutor())
    // 定义异步 I/O 回调函数
    override def asyncInvoke(str: String, resultFuture: ResultFuture[(String, String)]):
Unit = {
        val resultFutureRequested: Future[String] = client.query(str)
        resultFutureRequested.onSuccess {
            case result: String => resultFuture.complete(Iterable((str, result)))
        }
    }
}

val stream: DataStream[String] = ...
val resultStream: DataStream[(String, String)] =
AsyncDataStream.unorderedWait(stream, new AsyncDatabaseRequest(),
                    1000, TimeUnit.MILLISECONDS, 100)
```

3.6 状态管理

3.6.1 状态分类

状态分为如下两种类型。

（1）Keyed State，即定义在 KeyedStream 上的函数和 Operator 的状态。每一个 Operator 通常会有多个并行实例，但相同 key 的数据只能由同一个实例处理，因此一个 Keyed 状态只会对应一个 Operator 实例，但一个 Operator 实例会有多个状态分区。

（2）Non-Keyed State，即非分区的 Operator 状态。Kafka 连接器的每一个并行实例只负责一个消息分区，因此对应消息消费位置偏移就是这个连接器实例的非分区状态。

每一类状态有以下两种托管方式。

（1）托管方式（Managed State）。这类状态的数据结构由引擎定义，Flink 运行时负责序列化及写入状态后端。当并行度改变时，Flink 引擎负责重新拆分托管状态到各实例上。此外，引擎负责优化托管状态的存储效率。

（2）非托管方式（Raw State）。这类状态由应用程序定义，引擎以字节流格式写入状态后端。

3.6.2 托管的 Keyed State

Flink 内置了几种托管的 Keyed State。

- ValueState<T>：状态是单值的。
- ListState<T>：状态是多值的。
- ReducingState<T>：Reduce 函数的状态。
- AggregatingState<IN, OUT>：聚合函数的状态。

- MapState<UK, UV>：Map 函数的状态。

Flink 通过状态描述符（StateDescriptor）管理状态，以下是 ValueState 的状态描述符实现类：

```java
public class ValueStateDescriptor<T> extends StateDescriptor<ValueState<T>, T> {
    public ValueStateDescriptor(String name, Class<T> typeClass) {
        super(name, typeClass, null);
    }
    // 通过名称和状态的数据类型初始化描述符
    public ValueStateDescriptor(String name, TypeInformation<T> typeInfo) {
        super(name, typeInfo, null);
    }
    public ValueStateDescriptor(String name, TypeSerializer<T> typeSerializer) {
        super(name, typeSerializer, null);
    }
    @Override
    public Type getType() {
        return Type.VALUE;
    }
}
```

下面以求和为例，编程实现托管 ValueState 状态：

```scala
// 定义 FlatMap 函数
class CountWindowAverage extends RichFlatMapFunction[(Long, Long), (Long, Long)] {
  // ValueState 的数据类型为 (Long, Long) 元组
  private var sum: ValueState[(Long, Long)] = _
  override def flatMap(input: (Long, Long), out: Collector[(Long, Long)]): Unit = {
    // 读取状态值
    val tmpCurrentSum = sum.value
    val currentSum = if (tmpCurrentSum != null) {
      tmpCurrentSum
    } else {
      (0L, 0L)
    }
    val newSum = (currentSum._1 + 1, currentSum._2 + input._2)
    // 更新状态
    sum.update(newSum)
    // 求和、计数，当数量为 2 时输出中间结果并清空状态
    if (newSum._1 >= 2) {
      out.collect((input._1, newSum._2))
```

```
    sum.clear()
  }
}
// 通过名称和状态的数据类型初始化描述符
override def open(parameters: Configuration): Unit = {
  sum = getRuntimeContext.getState(
    new ValueStateDescriptor[(Long, Long)]("average", createTypeInformation[(Long, Long)])
  )
}
}

object ExampleCountWindowAverage extends App {
  val env = StreamExecutionEnvironment.getExecutionEnvironment
  env.fromCollection(List(
    (1L, 3L),
    (1L, 5L),
    (1L, 7L),
    (1L, 4L),
    (1L, 2L)
  )).keyBy(_._1)
    .flatMap(new CountWindowAverage())
    .print()
  env.execute("ExampleManagedState")
}
```

以上代码的输出为(1, 8), (1, 11)。如果将输入改为(2L, 3L), (2L, 5L), (2L, 7L), (1L, 4L), (1L, 2L), 则输出为(2,8), (1,6)。

3.6.3 状态后端配置

状态后端可通过 Flink 配置文件配置, 也可以在每个 Job 中单独配置。在 Job 中的配置方式如下:

```
val env = StreamExecutionEnvironment.getExecutionEnvironment()
env.setStateBackend(...)
```

其中, setStateBackend 的参数为状态后端, 有以下三种:

(1) 内存(MemoryStateBackend), 如 JVM 堆内存、JVM 堆外存, 设置状态

后端的代码如下：

```
val env = StreamExecutionEnvironment.getExecutionEnvironment()
env.setStateBackend(new MemoryStateBackend(MAX_MEM_STATE_SIZE, true))
```

（2）分布式文件系统（FsStateBackend），设置状态后端的代码如下：

```
val env = StreamExecutionEnvironment.getExecutionEnvironment()
env.setStateBackend(new FsStateBackend(path, true))
```

（3）RocksDB（RocksDBStateBackend），这种方式可以实现增量式状态。这种状态的后端配置为文件系统 URL，设置状态后端的代码如下：

```
val env = StreamExecutionEnvironment.getExecutionEnvironment()
env.setStateBackend(hdfs://namenode:40010/flink/checkpoints)
```

3.7 检查点

检查点是 Flink 实现 exactly-once 语义的核心机制，启用检查点机制需要具备以下两个条件：

（1）支持时空穿梭的外部数据源，如 Kafka、分布式文件系统。

（2）可持久化状态的外部存储，如分布式文件系统。

检查点默认是关闭的，启用检查点需要配置如下参数：

（1）一致性级别，即 exactly-once 或者 at-least-once。

（2）检查点超时时间，即检查点应该在多长时间内完成。

（3）两个检查点之间的最小时间间隔。

（4）并发检查点数量。默认情况下，在一个检查点还在处理快照逻辑时，Flink 不会触发另一个检查点，以确保低延迟。

（5）持久化检查点到外部存储。

（6）快照失败后任务是继续正常执行还是失败。

以下是配置代码：

```scala
val env = StreamExecutionEnvironment.getExecutionEnvironment()
// 启用检查点，设置两个检查点之间的最小时间间隔为 1000ms
env.enableCheckpointing(1000)

// 设置一致性级别为 exactly-once
env.getCheckpointConfig.setCheckpointingMode(CheckpointingMode.EXACTLY_ONCE)

// 设置检查点超时时间为 60000ms。如果在 60000ms 后还没有完成，则丢弃这个检查点
env.getCheckpointConfig.setCheckpointTimeout(60000)

// 设置快照失败后任务继续正常执行
env.getCheckpointConfig.setFailTasksOnCheckpointingErrors(false)

// 设置并发检查点数量为 1
env.getCheckpointConfig.setMaxConcurrentCheckpoints(1)
```

3.8 思考题

（1）我们在第 2 章中约定以 Scala 为编程语言，但为什么本章中很多代码用 Java 编写呢？

（2）在迟到生存期结束后窗口会被销毁，那么清除器的应用场景是什么呢？

（3）Source 和 Sink 也是计算图的一部分，如果对应的连接器不支持检查点机制会带来什么结果？

（4）Flink 是如何获取对象类型信息的？

（5）在设计增量式持久化状态时，Flink 架构需要考虑哪些问题？

第 4 章 批处理 API

批处理 API（DataSet API）和流处理 API 具有相同的编程规范，两者的应用程序结构基本相同，4.1 节将概括地介绍批处理的程序结构、Source、Sink 与连接器等内容；4.2 节将介绍算子，包括常见算子函数、广播变量、文件缓存、容错；4.3 节先介绍两个机器学习的例子，并从中总结出数据处理的通用特征，让读者更容易理解迭代应用场景。然后，介绍迭代的两种基本形式；4.4 节介绍注解，详细分析批处理程序的语义优化方法。

4.1 批处理 API 概述

批处理程序从 Source 拉取数据，通过 Pipeline 操纵 DataSet，并将结果写入 Sink。

由于所处理的对象是有界数据集，批处理不需要时间与窗口机制。因为所处理对象的结构特征是已知的，所以批处理程序可以引入新的技术，如迭代、注解，我们将在后续章节中深入讲解这些内容。

4.1.1 程序结构

批处理程序也分为 5 个部分。

1. 获取批处理运行时

```
// 批处理运行时
val env: ExecutionEnvironment = ExecutionEnvironment.getExecutionEnvironment
```

2. 添加 Source

可以添加外部数据源，也可以由应用创建 DataSet：

```
val text = env.fromElements(
  "Who's there?",
  "I think I hear them. Stand, ho! Who's there?")
```

3. 定义算子转换函数

下面的代码将输入 text 转换成小写形式、切分成单词、过滤空元素，然后映射成(word, 1)的数据结构，按照 word 分组，最后对每个分组计算单词出现的次数，输出结果为 DataSet[Int]：

```
val counts = text.flatMap { _.toLowerCase.split("\\W+") filter { _.nonEmpty } }
  .map { (_, 1) }
  .groupBy(0)
  .sum(1)
```

4. 定义 Sink

Sink 将处理结果写入外部系统，如打印到控制台：

```
// 打印到控制台
counts.print()
```

5. 启动程序

调用运行时的 execute()方法：

```
// 启动程序
env.execute("Kafka Dataset WordCount")
```

以上程序的完整版如下：

```scala
object WordCount {
  def main(args: Array[String]) {
    val env = ExecutionEnvironment.getExecutionEnvironment
    val text = env.fromElements(
      "Who's there?",
      "I think I hear them. Stand, ho! Who's there?")
    val counts = text.flatMap { _.toLowerCase.split("\\W+") filter { _.nonEmpty } }
      .map { (_, 1) }
      .groupBy(0)
      .sum(1)
    counts.print()
    // env.execute("Kafka Dataset WordCount")
  }
}
```

运行结果如下：

```
(hear,1)
(ho,1)
(i,2)
(stand,1)
(who,2)
(s,2)
(them,1)
(there,2)
(think,1)
```

4.1.2 Source

Source 包括消息队列、Socket 和文件，以及使用连接器连接的外部系统，接口定义如下：

1. 文件数据源

```
// 按行读取整个文件，path 代表文件路径
readTextFile(filePath: String)
// 按行读取 CSV 格式的文件，path 代表文件路径
readCsvFile(path)
// 定义文件格式 inputFormat
```

```
readFile (inputFormat: FileInputFormat[T], filePath: String)
```

2. 根据集合创建 DataStream

```
// 根据相同数据类型的元素创建 DataStream
fromElements (data: T*)
```

3. 递归地读取整个目录下的所有文件

```
// 设置递归读取
val parameters = new Configuration
parameters.setBoolean("recursive.file.enumeration", true)

// 递归读取
env.readTextFile("file:///path/with.nested/files").withParameters(parameters)
```

4.1.3 Sink

Sink 包括文件、控制台的外部系统，接口定义如下：

```
// 写入 csv 文件
writeAsCsv(path: String)
// 写入文本文件
writeAsText(path: String)
// 打印到控制台
print()
```

4.1.4 连接器

Source 和 Sink 可连接外部分布式文件系统，Flink 提供如下 4 类内置的分布式文件系统连接器。

（1）HDFS，以 hdfs://标识。

（2）Amazon S3，以 s3://标识。

（3）MapR，以 maprfs://标识。

（4）Alluxio，以 alluxio://标识。

4.2 算子

4.2.1 算子函数

1. Map

有以下三类 Map 操作。

（1）Map：对整个 DataSet 做一对一映射，即每一个元素产生一个输出。

（2）FlatMap：对整个 DataSet 做一对多映射，即每一个元素产生多个（也可以是一个）输出。

（3）MapPartition：Map 和 FlatMap 转换的对象是每一个元素，而 MapPartition 转换的对象是 DataSet 的每个分区。

```
//Map 操作
val intPairs: DataSet[(Int, Int)] = // [...]
val intSums = intPairs.map { pair => pair._1 + pair._2 }
//FlatMap 操作
val textLines: DataSet[String] = // [...]
val words = textLines.flatMap { _.split(" ") }

// MapPartition 中的 in 是一个 texLines 分区，结果是计算每个分区的大小
val textLines: DataSet[String] = // [...]
val counts = texLines.mapPartition { in => Some(in.size) }
```

2. Filter

过滤出条件为真的元素：

```
val intNumbers: DataSet[Int] = // [...]
// 过滤出值大于 0 的元素
val naturalNumbers = intNumbers.filter { _ > 0 }
```

3. SortPartition

根据数据的某个属性域升序或降序排序：

```scala
val tData: DataSet[(Int, String, Double)] = // [...]
val pData: DataSet[(BookPojo, Double)] = // [...]
val sData: DataSet[String] = // [...]

// 根据String属性升序排列
tData.sortPartition(1, Order.ASCENDING).print()

// 先根据Double属性降序排列，再根据Int属性升序排列
tData.sortPartition(2, Order.DESCENDING).sortPartition(0, Order.ASCENDING).print()

// 根据BookPojo的author属性降序排列
pData.sortPartition("_1.author", Order.DESCENDING).writeAsText(...)

// 根据所有属性升序排列
tData.sortPartition("_", Order.ASCENDING).writeAsCsv(...)
```

4．分组

数据处理中除 Map 外的另一个常见函数是 Reduce，即将多个元素映射为一个元素。在批处理程序中，这类函数可以作用于整个 DataSet，也可以作用于 DataSet 的每一个分组。批处理 API 没有 keyBy 函数，我们可以通过 groupBy 将 DataSet 分组，在每个分组上定义聚合函数，以下分情况展开分析。

1）Reduce

（1）作用于分组上。

```scala
// 定义WordCount类
class WC(val word: String, val count: Int) {
  def this() {
    this(null, -1)
  }
  // [...]
}

val words: DataSet[WC] = // [...]
// 根据word分组，求每个单词出现的次数
val wordCounts = words.groupBy("word").reduce {
  (w1, w2) => new WC(w1.word, w1.count + w2.count)
}
```

```
// 根据 KeySelector 分组
val wordCounts = words.groupBy { _.word } reduce {
  (w1, w2) => new WC(w1.word, w1.count + w2.count)
}

val tuples = DataSet[(String, Int, Double)] = // [...]
// 根据属性位置分组
val reducedTuples = tuples.groupBy(0, 1).reduce { ... }
```

（2）作用于整个 DataSet 上。

```
val intNumbers: DataSet[Int] = [...]
val sum = intNumbers.reduce (_ + _)
```

2）Aggregate

（1）作用于分组上。定义在同一个 DataSet 上的多个 Aggregate 函数之间用 and 连接，Aggregate 函数包括求和（SUM）、求最小值（MIN）和求最大值（MAX）。

```
// 根据 String 属性分组后，先对 Int 属性求和，再对 Double 属性求最小值
  val input: DataSet[(Int, String, Double)] = env.fromElements(
    (1, "Hello", 4),
    (1, "Hello", 5),
    (2, "Hello", 5),
    (3, "World", 6),
    (3, "World", 6)
  )
  val output = input.groupBy(1).
    aggregate(Aggregations.SUM, 0).
    and(Aggregations.MIN, 2)
```

以上程序的结果为：

```
(4,Hello,4.0)
(6,World,6.0)
```

（2）作用于整个 DataSet 上。

```
  val input: DataSet[(Int, String, Double)] = env.fromElements(
    (1, "Hello", 4.0),
    (1, "Hello", 5.0),
    (2, "Hello", 5.0),
    (3, "World", 6.0),
```

```
    (3, "World", 6.0)
)
val output = input.
  aggregate(Aggregations.SUM, 0).
  and(Aggregations.MIN, 2)
```

以上程序的结果为：

```
(10,World,4.0)
```

3）minBy/maxBy

（1）作用于分组上。

```
val input: DataSet[(Int, String, Double)] = env.fromElements(
  (1, "Hello", 5.0),
  (1, "Hello", 4.0),
  (2, "Hello", 5.0),
  (3, "World", 7.0),
  (4, "World", 6.0)
)
// 根据 String 属性分组后，先对 Int 属性求最小值，如果有多个相同 Int 的最小值，再对 Double 属性求最小值
val output = input.groupBy(1).
  minBy(0,2)
```

以上程序的结果为：

```
(1,Hello,4.0)
(3,World,7.0)
```

如果 minBy 的顺序调换，即

```
val output = input.groupBy(1).
  minBy(2,0)
```

则结果如下：

```
(1,Hello,4.0)
(4,World,6.0)
```

（2）作用于整个 DataSet 上。

```
val input: DataSet[(Int, String, Double)] = env.fromElements(
```

```
    (1, "Hello", 5.0),
    (1, "Hello", 4.0),
    (2, "Hello", 5.0),
    (3, "World", 7.0),
    (4, "World", 6.0)
)
val output = input.
  maxBy(2,0)
```

以上程序的结果为：

```
(3,World,7.0)
```

4）GroupReduce

（1）作用于分组上。与 Reduce 不同的是，GroupReduce 可以获取整个分组，通过迭代器访问分组的所有元素。

```
val input: DataSet[(Int, String, Double)] = env.fromElements(
    (1, "Hello", 5.0),
    (1, "Hello", 4.0),
    (2, "Hello", 5.0),
    (3, "World", 7.0),
    (4, "World", 6.0)
)
val output = input.groupBy(1).reduceGroup {
  (in, out: Collector[String]) =>
    out.collect(in.toSet.toString())
}
```

以上程序输出每一个分组的所有数据，结果如下：

```
Set((1,Hello,5.0), (1,Hello,4.0), (2,Hello,5.0))
Set((3,World,7.0), (4,World,6.0))
```

（2）作用于整个 DataSet 上。GroupReduce 获取整个 DataSet，通过迭代器访问 DataSet 的所有元素。

```
val input: DataSet[(Int, String, Double)] = env.fromElements(
    (1, "Hello", 5.0),
    (1, "Hello", 4.0),
    (2, "Hello", 5.0),
```

```
    (3, "World", 7.0),
    (4, "World", 6.0)
  )
  val output = input.reduceGroup {
    (in, out: Collector[String]) =>
      out.collect(in.toSet.toString())
  }
```

以上程序的结果为：

```
Set((4,World,6.0), (1,Hello,5.0), (1,Hello,4.0), (2,Hello,5.0), (3,World,7.0))
```

5. 去重

Flink 支持每个元素去重（Distinct），也支持根据 key 的位置去重：

```
// 对所有元素去重
val input: DataSet[(Int, String, Double)] = // [...]
val output = input.distinct()

// 对 Int 和 String 属性去重
val input: DataSet[(Int, Double, String)] = // [...]
val output = input.distinct(0,2)

// 根据转换结果去重
val input: DataSet[Int] = // [...]
val output = input.distinct {x => Math.abs(x)}
```

6. 连接

连接分为内连接和外连接，外链接又分为左外连接、右外连接和全外连接。内连接有以下 3 种形式。

（1）不带连接函数的形式。where 和 equalTo 分别定义被连接的两个数据集的属性位置，输出为这两个位置上相等的元素。

```
val input1: DataSet[(Int, String)] = env.fromElements((1,"hello"), (2,"world"))
val input2: DataSet[(String, Int)] = env.fromElements(("hello", 1), ("world", 2))
val result = input1.join(input2).where(0).equalTo(1)
```

以上程序的结果为：

```
((1,hello),(hello,1))
((2,world),(world,2))
```

（2）带 JoinFunction 连接函数的形式。

```
case class Rating(name: String, category: String, points: Int)
val ratings: DataSet[Rating] = env.fromElements(Rating("AA", "world", 10))
val weights: DataSet[(String, Double)] = env.fromElements(("hello", 1.0), ("world", 2.0))
val weightedRatings = ratings.join(weights).where("category").equalTo(0) {
    (rating, weight) => (rating.name, rating.points * weight._2)
}
```

以上程序的结果为：

```
(AA,20.0)
```

（3）带 FlatJoinFunction 连接函数的形式，返回一个 Collection。其与 JoinFunction 形式的关系可与 Map 和 FlatMap 的关系类比，下面的例子返回空集：

```
case class Rating(name: String, category: String, points: Int)
val ratings: DataSet[Rating] = env.fromElements(Rating("AA", "world", 10))
val weights: DataSet[(String, Double)] = env.fromElements(("hello", 1.0), ("world", 2.0))
val weightedRatings = ratings.join(weights).where("category").equalTo(0) {
 (rating, weight, out: Collector[(String, Double)]) =>
    if (weight._2 > 2) out.collect(rating.name, rating.points * weight._2)
}
```

7. 直积

笛卡儿积（Cartesian Product）是指在数学中，两个集合 X 和 Y 的笛卡儿积，又称直积，表示为 $X \times Y$。

```
case class Coord(id: Int, x: Int, y: Int)
val coords1: DataSet[Coord] = env.fromElements(Coord(5, 6, 7), Coord(8, 9, 10))
val coords2: DataSet[Coord] = env.fromElements(Coord(11, 12, 13),Coord(14, 15, 16))
val distances = coords1.cross(coords2) {
(c1, c2) =>
    val dist = math.sqrt(math.pow(c1.x - c2.x, 2) + math.pow(c1.y - c2.y, 2))
    (c1.id, c2.id, dist)
}
```

以上程序的结果为：

```
(5,11,8.48528137423857)
(5,14,12.727922061357855)
(8,11,4.242640687119285)
(8,14,8.48528137423857)
```

8. Union

类似关系数据库中的 Union，DataSet 的 Union 将多个相同类型的数据集拼接在一起。

```scala
val vals1: DataSet[(String, Int)] = env.fromElements(("hello", 1), ("world", 2))
val vals2: DataSet[(String, Int)] = env.fromElements(("hello", 1), ("world", 2))
val vals3: DataSet[(String, Int)] = env.fromElements(("hello", 1), ("world", 2))
val unioned = vals1.union(vals2).union(vals3)
```

以上程序的结果为：

```
(hello,1)
(hello,1)
(hello,1)
(world,2)
(world,2)
(world,2)
```

9. 数据分区

有以下 3 种数据分区模式。

（1）Rebalance 模式，根据轮询调度算法，将数据均匀地分发给下一级节点。

```scala
val in: DataSet[String] = // [...]
val out = in.rebalance().map { ... }
```

（2）Hash-Partition 模式，根据元组的某个属性域进行散列分区。

```scala
val in: DataSet[(String, Int)] = // [...]
val out = in.partitionByHash(0).mapPartition { ... }
```

（3）Range-Partition 模式，根据某个属性的范围进行分区。

```
val in: DataSet[(String, Int)] = // [...]
val out = in.partitionByRange(0).mapPartition { ... }
```

4.2.2 广播变量

广播变量（Broadcast Variable）是算子的多个并行实例间共享数据的一类方法，主要特点如下：

（1）动态数据共享。算子间共享的输入和配置参数是静态的，而广播变量共享的数据是动态的。

（2）可以分发更大规模的对象。应用程序可引用闭包（closure of function）中的变量，这种数据共享对象是小规模的，而广播变量则能分发更大规模的对象。

（3）广播变量以名称广播和访问（Access）。

广播变量以集合的方式定义在某个需要共享的算子上，算子的每一个实例可以通过集合访问共享变量。共享变量会在任务初始化时被发送到并行实例所在的节点上，并存储在 TaskManager 的内存里，因此尽管用于分发大规模的对象，共享变量也不易过大。

广播变量的编程步骤如下：

（1）创建广播变量。以下是通过 Collection 创造类型为 DataSet[Int]的广播变量的方法，这与普通 DataSet 的创建方法相同。

```
val toBroadcast = env.fromElements(1, 2, 3)
```

（2）注册广播变量。利用 RichFunction 实现自定义算子函数，在算子函数后注册广播变量，在算子函数的 open 方法中访问广播变量。

```
val data : DataSet[Int] = // [...]
data.map(new RichMapFunction[String, String]() {
    var broadcastSet: Traversable[String] = null
    override def open(config: Configuration): Unit = {
        // 读取广播变量
        broadcastSet=getRuntimeContext().getBroadcastVariable[String]("broadcastSet
```

```
Name").asScala
   }
   def map(in: String): String = {
     ...
   }
// 注册广播变量,名称为broadcastSetName
}).withBroadcastSet(toBroadcast, "broadcastSetName")
```

4.2.3 文件缓存

为了提高访问速度与效率,TaskManager 将算子实例要访问的远程文件复制到本地缓存起来,编程步骤如下。

第 1 步:注册缓存文件,文件分为两类。

(1)远程文件。这里的远程是相对于 JobManager 来说的,以下是注册 HDFS 缓存文件的编程示例。

```
val env = ExecutionEnvironment.getExecutionEnvironment
// 注册缓存文件,名称为hdfsFile
env.registerCachedFile("hdfs:///path/to/your/file", "hdfsFile")
```

(2)本地文件,即 JobManager 本地文件。

```
// 注册缓存文件,名称为localExecFile,最后一个参数定义是否为可执行文件
env.registerCachedFile("file:///path/to/exec/file", "localExecFile", true)
```

第 2 步:利用 RichFunction 实现自定义算子函数,通过注册名称访问缓存的文件或目录。

```
val env = ExecutionEnvironment.getExecutionEnvironment
class MyMapper extends RichMapFunction[String, Int] {
  override def open(config: Configuration): Unit = {
    // 获取缓存文件hdfsFile
    val myFile: File = getRuntimeContext.getDistributedCache.getFile("hdfsFile")
    // 读取文件内容
    ...
  }
  override def map(value: String): Int = {
    ...
```

```
    }
}
```

4.2.4 容错

批处理程序容错的方法是重试（Retry），重试有两个参数，即故障发生后最多重试次数和在故障发生后延迟多长时间才开始重试，配置方法如下：

```
val env = ExecutionEnvironment.getExecutionEnvironment()
// 重试次数
env.setNumberOfExecutionRetries(3)
// 重试延迟
env.getConfig.setExecutionRetryDelay(5000) // 5000 milliseconds delay
```

4.3 迭代

首先，以深度神经网络训练和网络社团发现算法为例，从中抽象出通用的数据处理特征，以帮助读者理解迭代的形式及其应用场景。

4.3.1 深度神经网络训练

人工神经网络（ANN，Artificial Neural Network）是人工智能的热门研究方向，是从人脑神经元网络处理信息抽象出的一种算法模型，通常简称为神经网络。这种模型由大量多层神经元节点相互连接构成，每两个节点间的连接代表通过该连接信号的加权，称之为权重。每个节点通过激励函数（Activation Function）将输入信号转换后输出到下一层神经元节点，如图 4-1 所示。

对于固定连接方式、层数、节点数的网络，通常采用误差逆传播算法（BP，error BackPropagation）训练神经网络得到权重和激活函数偏置，BP 算法的一般过程如下。

（1）根据损失函数计算训练样本正向通过神经网络的输出和样本标签之间的误差，算法根据这个步骤计算出的误差决定是否进行下一步的反向传播。这要求在训练开始前,初始化网络所有节点的参数,如权重和激活函数偏置,常用的初始化方法有随机值、利用非监督训练初始化等。

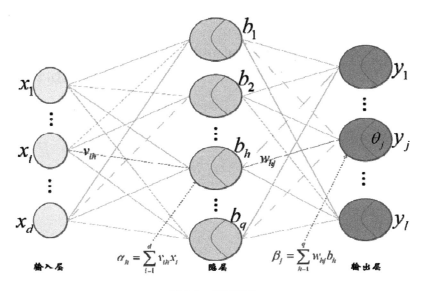

图 4-1 神经网络

(2)将第 1 步中计算的误差以一定的形式逐层回传到每一个神经元,后者根据回传的误差调整自己的参数,然后回到第 1 步。

因为 BP 算法是通过回传误差更新神经元参数的,所以常被称为 δ(delta)算法。

上述算法有两个数据处理特征。

(1)迭代计算:相同的训练样本集反复正向通过网络,经历相同的计算,这是迭代操作的特征之一。

(2)两个数据集:第一个数据集是训练样本(Work Set),其在这个算法中是不可变的,是每一次迭代的起点;第二个数据集是参数,每次正向迭代都参与计算,并且在每一次反向迭代时都会被更新,是可变的,也是算法的输出(Solution Set)。

批处理称以上迭代过程为 Delta Iteration。

4.3.2 网络社团发现算法

1998 年，Watts 和 Strogatz 在 *Nature* 上发表的复杂网络模型成功解释了"小世界现象"后，复杂网络模型成为诸多领域的基础模型，用于理解各个领域研究对象间复杂的拓扑关系和动力学行为。复杂网络的社团发现是复杂网络研究的一个重要方向，用于理解网络的拓扑结构、挖掘网络的潜在意义及预测网络行为等，因将分析所得的数据簇称为社团（community）而得名。社团发现算法的研究成果很丰富，标签传播算法（LPA，Label Propagation Algorithm）是其中之一。

每个节点被赋予一个特征标签，在每一次迭代计算过程中，节点的标签可能会被替换成与之相连的某个节点的标签。在经过若干次迭代后，彼此密集相连的节点会收敛于同一标签，最后具有相同标签的节点被归为一个社团。

在一个有两个连通域的无向图中，传播数值最小节点的特征，如图 4-2 所示。

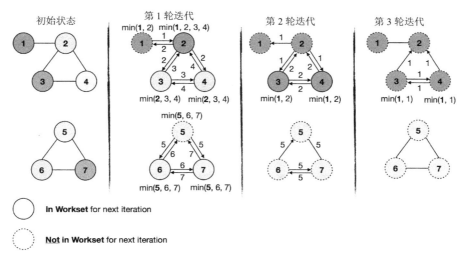

图 4-2 特征传播

（1）初始输入为两个数据集，每个数据集对应一个连通域，分别称为上连通域和下连通域。

（2）每一个元素都有自己的颜色特征，且各不相同。

（3）每一个元素对应图中的一个顶点，由三元组表示，即顶点 ID、顶点数值和顶点特征。其中顶点 ID 和顶点数值同为图中的数字。

现在的任务是在每一个连通区域内，用数字最小的顶点的特征替换其他顶点的特征，迭代计算的结束标志是每一个连通区域内所有顶点的特征达到一致。

在每一次迭代过程中，根据连接关系两两比较，并用数值小的顶点的特征替换数值大的顶点的特征，整个迭代过程如下。

（1）在第一次迭代中，ID 为 1 的顶点将特征传播给 ID 为 2 的顶点，ID 为 2 的顶点将原有特征传播给 ID 为 3 和 4 的顶点；ID 为 5 的顶点将特征传播给 ID 为 6 和 7 的顶点。ID 为 1 和 5 的顶点已经将自己的特征传播出去，因此不再参与后续迭代过程，即不再是 Work Set 的元素。

需要注意的是，这里提及的两两比较并不一定是最优的，本节的重点是分析迭代过程，因此不探讨最优的传播算法。

（2）在第二次迭代时下连通域的 Work Set 仅有 ID 为 6 和 7 的两个顶点，进行一轮特征传播就收敛了。由于此时这两个顶点当前的特征是被 ID 为 5 的顶点的特征替换后的特征，它们进行特征传播的过程并没有发生特征替换。

（3）第三次迭代将上连通域 ID 为 2 的顶点的特征，即初始状态 ID 为 1 的顶点的特征，传播给 ID 为 3 和 4 的顶点。

经过三次迭代，上下两个连通域都完成了特征传播。在这个过程中，Work Set 元素的数量是递减的，并在 Work Set 为空集时迭代结束，这在批处理中被称为 Bulk Iteration。这个例子与 BP 算法的不同在于其仅有一个数据集（即 Work Set），且是变化的。在分析完上述两个例子后，我们开始介绍批处理迭代的具体形式。

4.3.3 Bulk Iteration

Bulk Iteration 是一类比较简单的迭代形式，如图 4-3 所示。

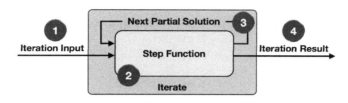

图 4-3　Bulk Iteration 的迭代形式

整个迭代过程描述如下：

（1）初始迭代输入（Iteration Input）可以来自 Source，也可以来自批处理上一级算子的输出。调用 DataSet 的迭代操作（Iterate）可以创建一个指定长度的可迭代数据集（IterativeDataSet）。

（2）迭代函数（Step Function）：每一轮迭代将调用迭代函数操作输入数据集，如 map、reduce 等。

（3）迭代输出（Next Partial Solution）：每一轮迭代将输入变换成另一个数据集。迭代将迭代输出反馈到迭代函数，作为下一轮迭代的输入，因此初始迭代输入仅作用于第一轮迭代。

（4）迭代结果（Iteration Result）：即迭代终止时的输出，因此必须定义迭代的终止条件，通常有以下两种形式。

- 最大迭代次数：如没有其他条件，迭代操作将在最大次数以内结束。这里的其他条件有多种形式，如特征传播例子中的 Work Set 为空集无法继续输入给迭代算子。
- 自定义收敛条件：应用程序可以自己实现聚合操作，收敛条件（Convergence Criterion）接口。

在可迭代数据集上调用终止条件操作，创建可迭代 DataSet，方法为 closeWith(DataSet, DataSet)，迭代在第二个参数为空集时结束。如果没有传入第二个参数，则根据给定的最大迭代次数确定结束点。

下面的例子是用蒙特卡罗方法计算π。先生成一个单元素 DataSet，对应为初始迭代输入：

```
// Create initial DataSet
val initial = env.fromElements(0)
```

然后设置最大迭代次数为 10000，并定义迭代函数为 map。在[0.0, 1.0)区间内随机生成一个坐标，如果此坐标在单位圆内，则将输入加 1 后发射出去：

```
val count = initial.iterate(10000) { iterationInput: DataSet[Int] =>
  val result = iterationInput.map { i =>
    val x = Math.random()
    val y = Math.random()
    i + (if (x * x + y * y < 1) 1 else 0)
  }
  result
}
```

迭代完成后，计算π值。需要注意的是，这里使用 map 函数进行转换，而不是对 Iteration Result 直接进行数学运算，因为程序会编译成并行计算图：

```
val result = count.map { c => c / 10000.0 * 4 }
result.print()
env.execute("Iterative Pi Example")
```

4.3.4 Delta Iteration 的迭代形式

Delta Iteration 的迭代形式，如图 4-4 所示。

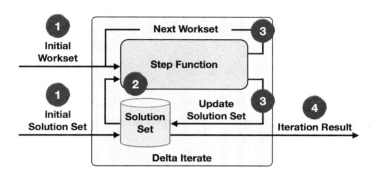

图 4-4　Delta Iteration 的迭代形式

整个迭代过程描述如下：

（1）初始迭代输入（Iteration Input）可以来自 Source，也可以来自批处理上一级算子的输出。

（2）迭代函数（Step Function）：每一轮迭代将调用迭代函数操纵输入数据集，如 map、reduce 等。

（3）迭代输出 WorkSet、更新 Solution Set：每一轮迭代将迭代函数作用于 WorkSet，并以迭代输出的 WorkSet 作为下一轮迭代的 WorkSet。此外，在每一轮迭代计算过程中，均可能更新迭代输入的 Solution Set。

（4）迭代结果（Iteration Result）：迭代终止时的输出 Solution Set。终止条件形式同 Bulk Iteration。

相比 Bulk Iteration，这种方法的效率通常更高，迭代仅更新 Solution Set 的热点部分（hot part），在结束前的几轮（或更多）迭代过程中，相当多数量的元素并不会发生变化。参考 BP 算法参数的调节过程就不难理解：如果在参数调节完成前仍有大量参数发生变化，通常是因为学习率设置得过大，这种神经网络训练结果的泛化效果不会太好。

在 Delta Iteration 编程时，需要先初始化 WorkSet 和 Solution Set，并确定最大迭代次数及 WorkSet 和 Solution Set 主键的位置：

```
val initialSolutionSet: DataSet[(Long, Double)] = // [...]
val initialWorkset: DataSet[(Long, Double)] = // [...]
val maxIterations = 100
val keyPosition = 0
```

然后，定义迭代函数，即根据当前的 WorkSet（workset）和 Solution Set（solution）计算下一轮的输入 WorkSet（nextWorkset）和 Solution Set（deltas）：

```
val result = initialSolutionSet.iterateDelta(initialWorkset, maxIterations,
Array(keyPosition)) {
  (solution, workset) =>
    val candidateUpdates = workset.groupBy(1).reduceGroup(new
ComputeCandidateChanges())
val deltas =
candidateUpdates.join(solution).where(0).equalTo(0)(new CompareChangesToCurrent())
    val nextWorkset = deltas.filter(new FilterByThreshold())
    (deltas, nextWorkset)
}
```

4.4 注解

Flink 批处理引擎可以借助数据处理逻辑优化应用程序的物理执行计划，如 DataSet 的哪些部分参与算子函数的计算；哪些部分未经更改地发射到下一级节点，注解就是定义这类数据处理逻辑的工具。需要注意的是，和实际不一致的注解通常会导致应用程序得出错误的计算结果，因此在没有完全清楚所有数据处理逻辑时，不建议使用注解。下面介绍注解的三种形式。

4.4.1 直接转发

直接转发的输入域（Forwarded Field）是指输入 DataSet 的某个属性未经算子修改而直接输出，可以是输入输出的相同位置，也可以是不同位置，有以下 3 种形式。

（1）相同位置直接转发。注解 "_2" 表示输入数据的第 3 个属性被算子直接转发到输出的第 3 个属性。

（2）不同位置直接转发。注解"_0->_2"表示输入数据的第 1 个属性的数据被直接转发到输出的第 3 个属性位置。此外，注解支持通配符"*"，标识整个输入或输出，如"_0->*"表示输入数据的第 1 个属性的数据被直接转发到输出，即输出只有一个属性。

（3）多个位置直接转发。注解"_0;_2->_1;_3->_2"表示输入数据的第 1 个属性的数据被直接转发到输出的第 1 个属性位置，输入数据的第 3 个属性的数据被直接转发到输出的第 2 个属性位置，输入数据的第 4 个属性的数据被直接转发到输出的第 3 个属性位置。

有以下两种方法标注注解：

```
// 直接标注在算子函数实现类上
@ForwardedFields("_1->_3")
class MyMap extends MapFunction[(Int, Int), (String, Int, Int)]{
  ...
}

// 使用withForwardedFields 标注
data.map(myMapFnc).withForwardedFields("_1->_3")
```

4.4.2 非直接转发

非直接转发的输入域（Non-Forwarded Field）指输入和输出某个相同位置的数据发生了变化，且其他位置的数据均未发生变化。因此，这类注解的语义更强，错误的注解将产生更严重的数据处理错误。

这类注解定义在算子函数上：

```
// 这个注解表示输入输出只有第 3 个属性位置发生变化
@NonForwardedFields("_2")
class MyMap extends MapFunction[(Int, Int), (Int, Int)]{
  def map(value: (Int, Int)): (Int, Int) = {
    return (value._1, value._2 / 2)
  }
}
```

4.4.3 触达

触达的输入域（Read Field）指输入 DataSet 的某个属性被算子读取并参与计算。这类注解的语义较强，需要列出所有触达的输入域：

```
@ReadFields("_1; _4")
class MyMap extends MapFunction[(Int, Int, Int, Int), (Int, Int)]{
  def map(value: (Int, Int, Int, Int)): (Int, Int) = {
  if (value._1 == 42) {
    return (value._1, value._2)
  } else {
    return (value._4 + 10, value._2)
  }
 }
}
```

4.5 思考题

（1）为什么 DataStream API 中没有注解？

（2）为什么 DataSet API 里没有 keyBy 算子？

（3）批处理架构引入迭代会面临哪些问题？

（4）在底层实现机制上，groupBy 和 keyBy 的区别是什么？

（5）相比检查点，重试则显得"简陋"，为什么批处理架构这么设计容错机制呢？

第 5 章
机器学习引擎架构与应用编程

5.1 节以多项式曲线拟合为例讲解 FlinkML 应用编程结构；5.2 节在总结机器学习架构所面临的架构问题及 Scikit-learn 架构经验后，详细分析 FlinkML 的机器学习流水线实现机制；5.3 节深入分析多项式曲线拟合的实现机制；5.4 节介绍分类算法思想、SVM 的 FlinkML 实现及 SVM 在多分类任务中的应用编程；5.5 节介绍推荐算法的原理与应用。

5.1 概述

FlinkML 致力于设计一套适用于不同数据规模的可伸缩系统，减少胶水代码以降低不同机器学习组件间的黏合成本，迁移于 Scikit-learn 的机器学习编程模型可以帮助开发人员快速上手。此外，Pipeline、迭代等组件让 FlinkML 能够满足复杂机器学习任务的要求。

FlinkML 的规划包括以下内容：

- Pipeline
- 数据预处理
- 模型选择和性能评估

- 监督学习
- 非监督学习
- 推荐
- 文本分析
- 统计
- 分布式线性代数
- 实时机器学习

FlinkML 基于 Scala 语言实现了流式机器学习底层框架，我们很容易在此基础上进行扩展，以满足不同领域的机器学习应用。本章不仅讲述 FlinkML 的应用，还分析底层架构实现及架构背后的理论。

FlinkML 是以库的形式提供的，在 Maven 工程中需要加入以下依赖：

```xml
<properties>
    <scala.flinkml.version>1.7.0</scala.flinkml.version>
    <scala.binary.version>2.11</scala.binary.version>
</properties>

<dependency>
    <groupId>org.apache.flink</groupId>
    <artifactId>flink-ml_${scala.binary.version}</artifactId>
    <version>${scala.flinkml.version}</version>
    <scope>provided</scope>
</dependency>
```

首先，我们以多项式曲线拟合为例展开分析。

5.1.1 数据加载

FlinkML 提供两种数据加载方式：ETL 方式和 LibSVM 数据格式加载方式。在监督学习任务中，LibSVM 是一个比较好的选择。

LabeledVector 类定义一个训练样本，其定义如下：

```
case class LabeledVector(label: Double, vector: Vector)
```

其中 vector 代表特征，即 $(\boldsymbol{x}, \boldsymbol{y})$ 中的 $\boldsymbol{x}$。

开源训练数据通常以文本文件的方式存储，如 .csv 文件。以 Haberman's Survival Data Set 为例，这个文件每一行代表一个样本，每一个样本有三个特征和一个标签，每两个数字之间以逗号分隔：

```
val survival = env.readCsvFile[(String, String, String,
String)]("/path/to/haberman.data")
```

下面的代码将上述数据由元组的形式转换成 LabeledVector 格式：

```
import org.apache.flink.ml.common.LabeledVector
import org.apache.flink.ml.math.DenseVector

val survivalLV = survival
  .map{tuple =>
    val list = tuple.productIterator.toList
    val numList = list.map(_.asInstanceOf[String].toDouble)
    LabeledVector(numList(3), DenseVector(numList.take(3).toArray))
  }
```

对 LibSVM 格式的文件来说，加载显得比较简洁：

```
import org.apache.flink.ml.MLUtils

val astroTrainLibSVM: DataSet[LabeledVector] = MLUtils.readLibSVM(env,
"/path/to/svmguide1")
```

5.1.2 多项式曲线拟合的例子

先看线性拟合的情况。在 n 维欧式空间中有 m 个训练样本

$$D = \{(\boldsymbol{x}_1, y_1), (\boldsymbol{x}_2, y_2), ..., (\boldsymbol{x}_m, y_m)\}$$

其中

$$\boldsymbol{x}_i = (x_{i1}, x_{i2}, ..., x_{in})$$

任务是拟合一条直线使得偏差最小，即求得

$$f(x) = w^T x + b$$

其中

$$x = (x_1, x_2, ..., x_n)$$

基于均方误差最小化（最小二乘法）的求解，可以得到闭式解（非数值解），求得w和b。

在机器学习中将上述样本训练过程称为拟合（fit），拟合结果$f(x_i)$称为在样本点上的估计（Estimation）。

为了拟合更复杂的样本分布情况，有时需要拟合多项式曲线，即

$$f(x) = w_k^T x^k + ... + w_1^T x + b$$

为了观察得更清楚，我们假定

$$n = 2, k = 2$$

将上面的式子展开后得

$$f(x) = a_1 x_1^2 + a_2 x_2^2 + a_3 x_1 x_2 + a_4 x_1 + a_5 x_2 + a_6$$

对参数的求解来说，上面的式子是线性的。因此，多项式曲线拟合过程的第一步是将样本空间变换到特征空间，即将x映射到

$$(x_1^2, x_2^2, x_1 x_2, x_1, x_2)$$

这种映射的本质是转换数据的形式；第二步是在特征空间中进行线性拟合。在实际的训练过程中，通常先对数据进行伸缩变换（scaler），如归一化、调整数据特征的分布形式等。

在 FlinkML 中实现上述拟合。首先，加载训练数据集和测试数据集：

```
val trainingData: DataSet[LabeledVector] = ...
val testingData: DataSet[Vector] = ...
```

创建 scaler、特征变换、多元线性回归：

```
val scaler = StandardScaler()
val polyFeatures = PolynomialFeatures().setDegree(2)
val mlr = MultipleLinearRegression()
```

构建机器学习的 Pipeline，即将前述定义的 scaler、多项式特征变换、多元线性回归"集成"在一起：

```
val pipeline = scaler.chainTransformer(polyFeatures).chainPredictor(mlr)
```

拟合样本数据集，即模型训练：

```
pipeline.fit(trainingData)
```

作用于测试数据，即模型预测，以分析模型的预测效果：

```
val predictions: DataSet[LabeledVector] = pipeline.predict(testingData)
```

5.2 流水线

5.2.1 机器学习面临的架构问题

为了降低机器学习的应用难度，通用的机器学习框架都会对数据分析的所有方面进行复杂的抽象，以便快速构建复杂机器学习任务的应用程序，这往往会带来"副作用"。针对特殊应用领域的软件开发，机器学习框架（引擎）开发面临巨大的挑战，重构、增加单元测试覆盖率、解耦等传统软件架构方法往往不能从容应对这些挑战。

1. 功能边界抽象困难

从样例中学习是机器学习的核心方法，这导致引擎的行为和数据紧密地耦合在一起，软件功能边界的抽象，如封装、模块解耦等工程方法，通常起不到决定性的作用。这从机器学习中的"没有免费的午餐"定理（No Free Lunch Theorem）可以看出端倪：每一个机器学习算法都不能解决所有领域的问题，只能解决与数据相关的、特定的、较窄的应用场景下的问题，因此引擎的结果（输出的训练模型）往往是与数据相关的。

而数据本身的相关性加剧了这种耦合。数据的多个属性间通常存在关联，这也是数据处理的重要问题之一，如特征选择。这种关联带来的后果是数据的微小变化导致训练结果差别很大（CACE，Changing Anything Changes Everything），而且这也适用于模型的参数调节。不仅如此，训练数据还会存在隐含的关联闭环。在机器学习中，通常提供模型的可视化工具来感性地观察这种关联，如高维可视化、指标度量（Metrics）等；正则化，如同系统的负反馈，也可以降低这种蝴蝶效应，但是还远远不够。

2. 过多胶水代码带来难以预知的问题

出于研究的目的，一些工程师会开发一套自包含的开源算法库，如支持向量机的 LIBSVM、深度学习的开源框架 TensorFlow 等，这导致引擎需要过多的代码去黏合不同的开源算法库，进而带来一些不可预知的系统问题。即便是同一引擎内部，为了将多个数据处理过程加载到同一训练数据集上，也需要考虑这些过程的黏合。为了减少应用层面的胶水代码，引擎通常提供系统级的黏合方案，如 Pipeline。

5.2.2　Scikit-learn 架构实践总结

Scikit-learn 是用 Python 语言实现的开源机器学习算法库，包括通用的机器学习算法、模型评估与选择工具、数据预处理。Scikit-learn 始于 2007 年的 Google Summer of Code 项目，最初由 David Cournapeau 开发。

Scikit-learn 的算法库建立在 SciPy（Scientific Python）之上，SciPy 的扩展和模块被命名为 SciKits，而提供学习算法的模组被命名为 Scikit-learn，主要包括以下几个部分。

- NumPy：基础的多维数组包。
- SciPy：科学计算的基础库。

- Matplotlib：全面的 2D 与 3D 测绘。
- IPython：改进的交互控制器。
- Sympy：符号数学。
- Pandas：数据结构和分析。

1. 底层数据结构

用 NumPy 多维数组表示稠密矩阵，用 SciPy 表示稀疏矩阵。这样分开表示是因为很多 Scikit-learn 用户已经习惯于 NumPy 和 SciPy 的数据表示方式，而且这也是在权衡计算性能的基础上做出的选择。数据处理或模型训练都是以批处理的方式进行的，一次输入一批数据而不是单个数据。Scikit-learn 针对批数据做了优化，而没有对单个数据做优化。

2. 估计

定义了机器学习对象的实例化机制，并暴露拟合方法，这两个机制的分离类似分段函数：初始化阶段只是将参数映射到机器学习算法，且这种映射发生在训练数据 feed（将数据输入训练模型的动作被称为 feed）之前；模型的训练则是通过调用拟合方法。除机器学习算法实现的，预处理、特征抽取等环节尽管没有拟合过程，其实现方式也是继承于 Estimator，FlinkML 沿用了这种实现哲学。

3. 预测

预测（Predictor）也是继承于 Estimator，在训练完成的算法上加上预测接口是很自然的想法，而性能评估则要求 Predictor 实现 score 方法。

4. 转换

Transformer 在 Estimator 的基础上增加了转换接口，变换输入数据，如数据归一化、调整数据概率分布等。

5. 流水线

Scikit-learn 提供将多个 Estimator 组合成一个新的 Estimator 对象的机制，这种机制就是流水线，在面向对象语言里用多继承来描述。流水线将多个机器学习工作流链接成一个超级 Estimator，其外在表现像单一 Estimator。从全局的角度展开参数优化（如模型选择），正是利用了流水线的机制。此外，Scikit-learn 提供的 FeatureUnion 机制也可以完成流水线功能，FlinkML 并没有借用这种机制。

6. 模型选择（Model Selection）

在决策树算法中，树的最大深度太小往往导致欠拟合（underfit），而太大则会导致过拟合（overfit），因此需要根据要解决的问题（训练数据集），通过算法决定最优超参（hyper-parameters）。需要注意的是，这类参数不是我们在线性回归中的参数 w 和 b，前者决定模型的某种具体形式，如支持向量机中软间隔参数 C，后者是在超参确定后进行模型训练确定的参数（详细情况参见 5.4 节）。对模型的选择通常采用网格搜索的方法，即在一系列离散的超参点上搜索最优解。

5.2.3 FlinkML 实现

1. ParameterMap

ParameterMap 定义了一个可序列化的、用于存储模型参数的可变 map。首先，定义无参数初始化方法：

```
class ParameterMap(val map: mutable.Map[Parameter[_], Any]) extends Serializable {
  def this() = {
    this(new mutable.HashMap[Parameter[_], Any]())
  }
```

Parameter 定义在同一个文件中，设置默认参数值，Option[T] 是一个类型为 T 的可选值的容器：

```
trait Parameter[T] {
  val defaultValue: Option[T]
}
```

然后定义了 map 的 "+" 操作符（插入元素）、重载了 map 的 "++" 运算符（合并）：

```
def add[T](parameter: Parameter[T], value: T): ParameterMap = {
  map += (parameter -> value)
  this
}
def ++(parameters: ParameterMap): ParameterMap = {
  val result = new ParameterMap(map)
  result.map ++= parameters.map
  result
}
```

再定义 map 的 get 操作,根据 map 中是否存在对应 key、默认参数及返回值类型,分三种情况定义 get 的返回值,FlinkML 将它们拆分成两个方法:

```
def get[T](parameter: Parameter[T]): Option[T] = {
  if(map.isDefinedAt(parameter)) {
    map.get(parameter).asInstanceOf[Option[T]]
  } else {
    parameter.defaultValue
  }
}
def apply[T](parameter: Parameter[T]): T = {
  if(map.isDefinedAt(parameter)) {
    map(parameter).asInstanceOf[T]
  } else {
    parameter.defaultValue match {
      case Some(value) => value
      case None => throw new NoSuchElementException(s"Could not retrieve " +
        s"parameter value $parameter.")
    }
  }
}
```

最后创建伴生对象:

```
object ParameterMap {
  val Empty = new ParameterMap
  def apply(): ParameterMap = {
    new ParameterMap
  }
}
```

在 Scala 语言中没有 static 这种语法,但提供了单例模式的实现方法,就是上面的关键字 object。

在 Scala 语言中使用单例模式时，除了要定义类，还要定义一个同名的 object 对象，但 object 对象不能带参数。单例对象分为两种，一种是并未自动关联到特定类上的单例对象，被称为独立对象（Standalone Object）；另一种是与类同名的单例对象，被称为是这个类的伴生对象（Companion Object）。

模型通过继承 WithParameters 管理配置参数：

```
trait WithParameters {
  val parameters = new ParameterMap
}
```

完整的 ParameterMap 代码如下：

```
/**
 * Map use d to store configuration parameters for algorithms. The parameter
 * * values are stored in a [[Map]] being identified by a [[Parameter]] object. ParameterMaps can
 * * be fused. This operation is left associative, meaning that latter ParameterMaps can override
 * * parameter values defined in a preceding ParameterMap.
 * *
 *
 * @param map Map containing parameter settings
 */
class ParameterMap(val map: mutable.Map[Parameter[_], Any]) extends Serializable {

  def this() = {
    this(new mutable.HashMap[Parameter[_], Any]())
  }

  /**
   * Adds a new parameter value to the ParameterMap.
   *
   * @param parameter Key
   * @param value Value associated with the given key
   * @tparam T Type of value
   */
  def add[T](parameter: Parameter[T], value: T): ParameterMap = {
    map += (parameter -> value)
```

```
    this
  }

  /**
    * Retrieves a parameter value associated to a given key. The value is returned as an Option.
    * If there is no value associated to the given key, then the default value of the [[Parameter]]
    * is returned.
    *
    * @param parameter Key
    * @tparam T Type of the value to retrieve
    * @return Some(value) if an value is associated to the given key, otherwise the default value
    *         defined by parameter
    */
  def get[T](parameter: Parameter[T]): Option[T] = {
    if(map.isDefinedAt(parameter)) {
      map.get(parameter).asInstanceOf[Option[T]]
    } else {
      parameter.defaultValue
    }
  }

  /**
    * Retrieves a parameter value associated to a given key. If there is no value contained in the
    * map, then the default value of the [[Parameter]] is checked. If the default value is defined,
    * then it is returned. If the default is undefined, then a [[NoSuchElementException]] is thrown.
    *
    * @param parameter Key
    * @tparam T Type of value
    * @return Value associated with the given key or its default value
    */
  def apply[T](parameter: Parameter[T]): T = {
    if(map.isDefinedAt(parameter)) {
      map(parameter).asInstanceOf[T]
    } else {
      parameter.defaultValue match {
        case Some(value) => value
        case None => throw new NoSuchElementException(s"Could not retrieve " +
```

```
            s"parameter value $parameter.")
      }
    }
  }

  /**
   * Adds the parameter values contained in parameters to itself.
   *
   * @param parameters [[ParameterMap]] containing the parameter values to be added
   * @return this after inserting the parameter values from parameters
   */
  def ++(parameters: ParameterMap): ParameterMap = {
    val result = new ParameterMap(map)
    result.map ++= parameters.map

    result
  }
}

object ParameterMap {
  val Empty = new ParameterMap

  def apply(): ParameterMap = {
    new ParameterMap
  }
}

/**
 * Base trait for parameter keys
 *
 * @tparam T Type of parameter value associated to this parameter key
 */
trait Parameter[T] {

  /**
   * Default value of parameter. If no such value exists, then returns [[None]]
   */
  val defaultValue: Option[T]
}
```

2. Estimator

Estimator 是 Pipeline 中所有对象的父类，实现了顶层 fit 方法。在其伴生对象中提供了 4 种 fallback 实现，当 fit、transform、predict 或 evaluate（用于模型泛化性能评估）调用出现异常时，这 4 种 fallback 实现将会被调用。对每一种输入数据类型，Estimator 的继承类必须实现对应版本的 fit 等方法，fallback 实现方法相当于类型检查器。而且，当一个 Pipeline 中的各机器学习对象的连接方式不对时，fallback 方法也会被调用。

先定义 fit 特质：

```
trait FitOperation[Self, Training]{
  def fit(instance: Self, fitParameters: ParameterMap, input: DataSet[Training]): Unit
}
```

再定义 Estimator 特质，其中 Training 是训练样本的数据类型，默认模型参数为空（在 ParameterMap 伴生对象中定义）。Estimator 继承类需要为每一种数据的类型对应实现 fit 方法，这里将 FitOperation 方法作为参数传入 fit 接口：

```
trait Estimator[Self] extends WithParameters {
 that: Self =>

 def fit[Training](
    training: DataSet[Training],
    fitParameters: ParameterMap = ParameterMap.Empty)(implicit
    fitOperation: FitOperation[Self, Training]): Unit = {
  FlinkMLTools.registerFlinkMLTypes(training.getExecutionEnvironment)
  fitOperation.fit(this, fitParameters, training)
 }
}
```

此外，fit 方法将向运行时环境注册 FlinkML 支持的数据类型，序列化需要类型信息：

```
FlinkMLTools.registerFlinkMLTypes(training.getExecutionEnvironment)
->
 def registerFlinkMLTypes(env: ExecutionEnvironment): Unit = {
  // Vector types
  env.registerType(classOf[org.apache.flink.ml.math.DenseVector])
```

```
env.registerType(classOf[org.apache.flink.ml.math.SparseVector])
// Matrix types
env.registerType(classOf[org.apache.flink.ml.math.DenseMatrix])
env.registerType(classOf[org.apache.flink.ml.math.SparseMatrix])
// Breeze Vector types
env.registerType(classOf[breeze.linalg.DenseVector[_]])
env.registerType(classOf[breeze.linalg.SparseVector[_]])
// Breeze specialized types
env.registerType(breeze.linalg.DenseVector.zeros[Double](0).getClass)
env.registerType(breeze.linalg.SparseVector.zeros[Double](0).getClass)
// Breeze Matrix types
env.registerType(classOf[breeze.linalg.DenseMatrix[Double]])
env.registerType(classOf[breeze.linalg.CSCMatrix[Double]])
// Breeze specialized types
env.registerType(breeze.linalg.DenseMatrix.zeros[Double](0, 0).getClass)
env.registerType(breeze.linalg.CSCMatrix.zeros[Double](0, 0).getClass)
}
```

以下是使用拟合方法的 fallback 抛出的异常样式：

```
implicit def fallbackFitOperation[
    Self: TypeTag,
    Training: TypeTag]
  : FitOperation[Self, Training] = {
  new FitOperation[Self, Training]{
    override def fit(
        instance: Self,
        fitParameters: ParameterMap,
        input: DataSet[Training])
      : Unit = {
      val self = typeOf[Self]
      val training = typeOf[Training]
      throw new RuntimeException("There is no FitOperation defined for " + self +
        " which trains on a DataSet[" + training + "]")
    }
  }
}
```

完整的 Estimator 代码如下：

```
/** Base trait for Flink's pipeline operators.
  *
  * An estimator can be fitted to input data. In order to do that the implementing class
has
```

```
 * to provide an implementation of a [[FitOperation]] with the correct input type. In
order to make
 * the [[FitOperation]] retrievable by the Scala compiler, the implementation should
be placed
 * in the companion object of the implementing class.
 *
 * The pipeline mechanism has been inspired by scikit-learn
 *
 * @tparam Self
 */
trait Estimator[Self] extends WithParameters {
  that: Self =>

  /** Fits the estimator to the given input data. The fitting logic is contained in the
    * [[FitOperation]]. The computed state will be stored in the implementing class.
    *
    * @param training Training data
    * @param fitParameters Additional parameters for the [[FitOperation]]
    * @param fitOperation [[FitOperation]] which encapsulates the algorithm logic
    * @tparam Training Type of the training data
    * @return
    */
  def fit[Training](
      training: DataSet[Training],
      fitParameters: ParameterMap = ParameterMap.Empty)(implicit
      fitOperation: FitOperation[Self, Training]): Unit = {
    FlinkMLTools.registerFlinkMLTypes(training.getExecutionEnvironment)
    fitOperation.fit(this, fitParameters, training)
  }
}

object Estimator{

  /** Fallback [[FitOperation]] type class implementation which is used if no other
    * [[FitOperation]] with the right input types could be found in the scope of the
implementing
    * class. The fallback [[FitOperation]] makes the system fail in the pre-flight phase
by
    * throwing a [[RuntimeException]] which states the reason for the failure. Usually
the error
    * is a missing [[FitOperation]] implementation for the input types or the wrong chaining
    * of pipeline operators which have incompatible input/output types.
    *
```

```
 * @tparam Self Type of the pipeline operator
 * @tparam Training Type of training data
 * @return
 */
implicit def fallbackFitOperation[
    Self: TypeTag,
    Training: TypeTag]
  : FitOperation[Self, Training] = {
  new FitOperation[Self, Training]{
    override def fit(
        instance: Self,
        fitParameters: ParameterMap,
        input: DataSet[Training])
      : Unit = {
      val self = typeOf[Self]
      val training = typeOf[Training]

      throw new RuntimeException("There is no FitOperation defined for " + self +
" which trains on a DataSet[" + training + "]")
    }
  }
}

/** Fallback [[PredictDataSetOperation]] if a [[Predictor]] is called with a not supported input.
  * data type. The fallback [[PredictDataSetOperation]] lets the system fail with a
  * [[RuntimeException]] stating which input and output data types were inferred but for which no
  * [[PredictDataSetOperation]] could be found.
  *
  * @tparam Self Type of the [[Predictor]]
  * @tparam Testing Type of the testing data
  * @return
  */
implicit def fallbackPredictOperation[
    Self: TypeTag,
    Testing: TypeTag]
  : PredictDataSetOperation[Self, Testing, Any] = {
  new PredictDataSetOperation[Self, Testing, Any] {
    override def predictDataSet(
        instance: Self,
        predictParameters: ParameterMap,
        input: DataSet[Testing])
```

```scala
        : DataSet[Any] = {
      val self = typeOf[Self]
      val testing = typeOf[Testing]

      throw new RuntimeException("There is no PredictOperation defined for " + self +
" which takes a DataSet[" + testing + "] as input.")
    }
  }
}

/** Fallback [[TransformDataSetOperation]] for [[Transformer]] which do not support the input or
  * output type with which they are called. This is usually the case if pipeline operators are
  * chained which have incompatible input/output types. In order to detect these failures, the
  * fallback [[TransformDataSetOperation]] throws a [[RuntimeException]] with the corresponding
  * input/output types. Consequently, a wrong pipeline will be detected at pre-flight phase of
  * Flink and thus prior to execution time.
  *
  * @tparam Self Type of the [[Transformer]] for which the [[TransformDataSetOperation]] is
  *              defined
  * @tparam IN Input data type of the [[TransformDataSetOperation]]
  * @return
  */
implicit def fallbackTransformOperation[
Self: TypeTag,
IN: TypeTag]
: TransformDataSetOperation[Self, IN, Any] = {
  new TransformDataSetOperation[Self, IN, Any] {
    override def transformDataSet(
      instance: Self,
      transformParameters: ParameterMap,
      input: DataSet[IN])
    : DataSet[Any] = {
      val self = typeOf[Self]
      val in = typeOf[IN]

      throw new RuntimeException("There is no TransformOperation defined for " +
        self + " which takes a DataSet[" + in +
```

```
"] as input.")
    }
  }
}

implicit def fallbackEvaluateOperation[
    Self: TypeTag,
    Testing: TypeTag]
  : EvaluateDataSetOperation[Self, Testing, Any] = {
  new EvaluateDataSetOperation[Self, Testing, Any] {
    override def evaluateDataSet(
      instance: Self,
      predictParameters: ParameterMap,
      input: DataSet[Testing])
    : DataSet[(Any, Any)] = {
    val self = typeOf[Self]
    val testing = typeOf[Testing]

    throw new RuntimeException("There is no PredictOperation defined for " + self +
" which takes a DataSet[" + testing + "] as input.")
    }
  }
 }
}

/** Type class for the fit operation of an [[Estimator]].
 *
 * The [[FitOperation]] contains a self type parameter so that the Scala compiler looks into
 * the companion object of this class to find implicit values.
 *
 * @tparam Self Type of the [[Estimator]] subclass for which the [[FitOperation]] is defined
 * @tparam Training Type of the training data
 */
trait FitOperation[Self, Training]{
  def fit(instance: Self, fitParameters: ParameterMap, input: DataSet[Training]): Unit
}
```

3. Predictor

类似 Estimator，Predictor 先定义 predict 和 evaluate 两种方法的原型。由于在

整个测试集上进行性能评估才有意义,在实际引用中需要预测单个样本的标签。相应地,predict 方法有两种类型,而 evaluate 只有一种形式。

以下是在数据集上的 predict 原型,只定义输入和输出的类型:

```
trait PredictDataSetOperation[Self, Testing, Prediction] extends Serializable{
  def predictDataSet(
    instance: Self,
    predictParameters: ParameterMap,
    input: DataSet[Testing])
  : DataSet[Prediction]
}
```

以下是单个样本的 predict 原型:

```
trait PredictOperation[Instance, Model, Testing, Prediction] extends Serializable{
  def getModel(instance: Instance, predictParameters: ParameterMap): DataSet[Model]
  def predict(value: Testing, model: Model):
    Prediction
}
```

这种 predict 原型考虑到更通用的操作,以 getModel 为例,所有训练的结果(如多项式曲线模型的参数 w 和 b)存储在 Pipeline 中(Instance),这提供了将这些参数输出的接口,因此 Model 就是模型参数集对象。

以下是 evaluate 的原型,其中输出结果为测试数据标签和预测标签组成的元组:

```
trait EvaluateDataSetOperation[Instance, Testing, Prediction] extends Serializable{
  def evaluateDataSet(
    instance: Instance,
    evaluateParameters: ParameterMap,
    testing: DataSet[Testing])
  : DataSet[(Prediction, Prediction)]
}
```

完成上述定义后,可以定义顶层 Predictor 的特质,其中 evaluate 调用的是在数据集上的接口:

```
trait Predictor[Self] extends Estimator[Self] with WithParameters {
```

```
that: Self =>

def predict[Testing, Prediction](
    testing: DataSet[Testing],
    predictParameters: ParameterMap = ParameterMap.Empty)(implicit
    predictor: PredictDataSetOperation[Self, Testing, Prediction])
  : DataSet[Prediction] = {
  FlinkMLTools.registerFlinkMLTypes(testing.getExecutionEnvironment)
  predictor.predictDataSet(this, predictParameters, testing)
}
def evaluate[Testing, PredictionValue](
    testing: DataSet[Testing],
    evaluateParameters: ParameterMap = ParameterMap.Empty)(implicit
    evaluator: EvaluateDataSetOperation[Self, Testing, PredictionValue])
  : DataSet[(PredictionValue, PredictionValue)] = {
  FlinkMLTools.registerFlinkMLTypes(testing.getExecutionEnvironment)
  evaluator.evaluateDataSet(this, evaluateParameters, testing)
}
}
```

按照 FlinkML 架构的"套路"，在 Predictor 的伴生对象中提供数据集上的 predict 和 evaluate 方法的默认实现。在实现过程中提供了模型参数的覆盖机制，即可以用 predictParameters 部分覆盖 instance 参数，这利用了运算符"++"的特性。此外，默认实现用到了单个元素的 predict 接口。有可能预测值与测试元素的数据类型并不相同，在实现中以 TypeInformation[PredictionValue]和 TypeInformation[Testing]标识二者的不同：

```
implicit def defaultPredictDataSetOperation[
    Instance <: Estimator[Instance],
    Model,
    Testing,
    PredictionValue](
    implicit predictOperation: PredictOperation[Instance, Model, Testing,
PredictionValue],
    testingTypeInformation: TypeInformation[Testing],
    predictionValueTypeInformation: TypeInformation[PredictionValue])
  : PredictDataSetOperation[Instance, Testing, (Testing, PredictionValue)] = {
  new PredictDataSetOperation[Instance, Testing, (Testing, PredictionValue)] {
    override def predictDataSet(
        instance: Instance,
```

```
      predictParameters: ParameterMap,
      input: DataSet[Testing])
    : DataSet[(Testing, PredictionValue)] = {
    val resultingParameters = instance.parameters ++ predictParameters
    val model = predictOperation.getModel(instance, resultingParameters)
    implicit val resultTypeInformation = createTypeInformation[(Testing,
PredictionValue)]
    input.mapWithBcVariable(model){
      (element, model) => {
        (element, predictOperation.predict(element, model))
      }
    }
  }
}
```

evaluate 的结果是测试数据标签和预测标签组成的元组：

```
implicit def defaultEvaluateDataSetOperation[
    Instance <: Estimator[Instance],
    Model,
    Testing,
    PredictionValue](
    implicit predictOperation: PredictOperation[Instance, Model, Testing,
PredictionValue],
    testingTypeInformation: TypeInformation[Testing],
    predictionValueTypeInformation: TypeInformation[PredictionValue])
  : EvaluateDataSetOperation[Instance, (Testing, PredictionValue), PredictionValue] =
{
  new EvaluateDataSetOperation[Instance, (Testing, PredictionValue), PredictionValue]
{
    override def evaluateDataSet(
        instance: Instance,
        evaluateParameters: ParameterMap,
        testing: DataSet[(Testing, PredictionValue)])
      : DataSet[(PredictionValue, PredictionValue)] = {
      val resultingParameters = instance.parameters ++ evaluateParameters
      val model = predictOperation.getModel(instance, resultingParameters)
      implicit val resultTypeInformation = createTypeInformation[(Testing,
PredictionValue)]
      testing.mapWithBcVariable(model){
        (element, model) => {
          (element._2, predictOperation.predict(element._1, model))
```

```
          }
        }
      }
    }
  }
}
```

完整的 Predictor 代码如下：

```
/** Predictor trait for Flink's pipeline operators.
  *
  * A [[Predictor]] calculates predictions for testing data based on the model it learned
during
  * the fit operation (training phase). In order to do that, the implementing class has
to provide
  * a [[FitOperation]] and a [[PredictDataSetOperation]] implementation for the correct
types. The
  * implicit values should be put into the scope of the companion object of the implementing
class
  * to make them retrievable for the Scala compiler.
  *
  * The pipeline mechanism has been inspired by scikit-learn
  *
  * @tparam Self Type of the implementing class
  */
trait Predictor[Self] extends Estimator[Self] with WithParameters {
  that: Self =>

  /** Predict testing data according the learned model. The implementing class has to
provide
    * a corresponding implementation of [[PredictDataSetOperation]] which contains the
prediction
    * logic.
    *
    * @param testing Testing data which shall be predicted
    * @param predictParameters Additional parameters for the prediction
    * @param predictor [[PredictDataSetOperation]] which encapsulates the prediction
logic
    * @tparam Testing Type of the testing data
    * @tparam Prediction Type of the prediction data
    * @return
    */
  def predict[Testing, Prediction](
```

```
    testing: DataSet[Testing],
    predictParameters: ParameterMap = ParameterMap.Empty)(implicit
    predictor: PredictDataSetOperation[Self, Testing, Prediction])
  : DataSet[Prediction] = {
  FlinkMLTools.registerFlinkMLTypes(testing.getExecutionEnvironment)
  predictor.predictDataSet(this, predictParameters, testing)
}

/** Evaluates the testing data by computing the prediction value and returning a pair of true
  * label value and prediction value. It is important that the implementation chooses a Testing
  * type from which it can extract the true label value.
  *
  * @param testing
  * @param evaluateParameters
  * @param evaluator
  * @tparam Testing
  * @tparam PredictionValue
  * @return
  */
def evaluate[Testing, PredictionValue](
    testing: DataSet[Testing],
    evaluateParameters: ParameterMap = ParameterMap.Empty)(implicit
    evaluator: EvaluateDataSetOperation[Self, Testing, PredictionValue])
  : DataSet[(PredictionValue, PredictionValue)] = {
  FlinkMLTools.registerFlinkMLTypes(testing.getExecutionEnvironment)
  evaluator.evaluateDataSet(this, evaluateParameters, testing)
}
}

object Predictor {

/** Default [[PredictDataSetOperation]] which takes a [[PredictOperation]] to calculate a tuple
  * of testing element and its prediction value.
  *
  * Note: We have to put the TypeInformation implicit values for Testing and PredictionValue after
  * the PredictOperation implicit parameter. Otherwise, if it's defined as a context bound, then
  * the Scala compiler does not find the implicit [[PredictOperation]] value.
  *
```

```
 * @param predictOperation
 * @param testingTypeInformation
 * @param predictionValueTypeInformation
 * @tparam Instance
 * @tparam Model
 * @tparam Testing
 * @tparam PredictionValue
 * @return
 */
implicit def defaultPredictDataSetOperation[
    Instance <: Estimator[Instance],
    Model,
    Testing,
    PredictionValue](
    implicit predictOperation: PredictOperation[Instance, Model, Testing,
PredictionValue],
    testingTypeInformation: TypeInformation[Testing],
    predictionValueTypeInformation: TypeInformation[PredictionValue])
  : PredictDataSetOperation[Instance, Testing, (Testing, PredictionValue)] = {
  new PredictDataSetOperation[Instance, Testing, (Testing, PredictionValue)] {
    override def predictDataSet(
        instance: Instance,
        predictParameters: ParameterMap,
        input: DataSet[Testing])
      : DataSet[(Testing, PredictionValue)] = {
      val resultingParameters = instance.parameters ++ predictParameters

      val model = predictOperation.getModel(instance, resultingParameters)

      implicit val resultTypeInformation = createTypeInformation[(Testing,
PredictionValue)]

      input.mapWithBcVariable(model){
        (element, model) => {
          (element, predictOperation.predict(element, model))
        }
      }
    }
  }
}

/** Default [[EvaluateDataSetOperation]] which takes a [[PredictOperation]] to
calculate a tuple
```

```
   * of true label value and predicted label value.
   *
   * Note: We have to put the TypeInformation implicit values for Testing and
PredictionValue after
   * the PredictOperation implicit parameter. Otherwise, if it's defined as a context
bound, then
   * the Scala compiler does not find the implicit [[PredictOperation]] value.
   *
   * @param predictOperation
   * @param testingTypeInformation
   * @param predictionValueTypeInformation
   * @tparam Instance
   * @tparam Model
   * @tparam Testing
   * @tparam PredictionValue
   * @return
   */
  implicit def defaultEvaluateDataSetOperation[
      Instance <: Estimator[Instance],
      Model,
      Testing,
      PredictionValue](
      implicit predictOperation: PredictOperation[Instance, Model, Testing,
PredictionValue],
      testingTypeInformation: TypeInformation[Testing],
      predictionValueTypeInformation: TypeInformation[PredictionValue])
    : EvaluateDataSetOperation[Instance, (Testing, PredictionValue), PredictionValue] =
{
    new EvaluateDataSetOperation[Instance, (Testing, PredictionValue), PredictionValue]
{
      override def evaluateDataSet(
          instance: Instance,
          evaluateParameters: ParameterMap,
          testing: DataSet[(Testing, PredictionValue)])
        : DataSet[(PredictionValue, PredictionValue)] = {
        val resultingParameters = instance.parameters ++ evaluateParameters
        val model = predictOperation.getModel(instance, resultingParameters)

        implicit val resultTypeInformation = createTypeInformation[(Testing,
PredictionValue)]

        testing.mapWithBcVariable(model){
          (element, model) => {
```

```
              (element._2, predictOperation.predict(element._1, model))
        }
      }
    }
   }
 }
}

/** Type class for the predict operation of [[Predictor]]. This predict operation works
on DataSets.
  *
  * [[Predictor]]s either have to implement this trait or the [[PredictOperation]] trait.
The
  * implementation has to be made available as an implicit value or function in the scope
of
  * their companion objects.
  *
  * The first type parameter is the type of the implementing [[Predictor]] class so that
the Scala
  * compiler includes the companion object of this class in the search scope for the implicit
  * values.
  *
  * @tparam Self Type of [[Predictor]] implementing class
  * @tparam Testing Type of testing data
  * @tparam Prediction Type of predicted data
  */
trait PredictDataSetOperation[Self, Testing, Prediction] extends Serializable{

  /** Calculates the predictions for all elements in the [[DataSet]] input
    *
    * @param instance The Predictor instance that we will use to make the predictions
    * @param predictParameters The parameters for the prediction
    * @param input The DataSet containing the unlabeled examples
    * @return
    */
  def predictDataSet(
                instance: Self,
                predictParameters: ParameterMap,
                input: DataSet[Testing])
    : DataSet[Prediction]
}

/** Type class for predict operation. It takes an element and the model and then computes
```

```
 the
  * prediction value for this element.
  *
  * It is sufficient for a [[Predictor]] to only implement this trait to support the evaluate and
  * predict method.
  *
  * @tparam Instance The concrete type of the [[Predictor]] that we will use for predictions
  * @tparam Model The representation of the predictive model for the algorithm, for example a
  *               Vector of weights
  * @tparam Testing The type of the example that we will use to make the predictions (input)
  * @tparam Prediction The type of the label that the prediction operation will produce (output)
  *
  */
trait PredictOperation[Instance, Model, Testing, Prediction] extends Serializable{

  /** Defines how to retrieve the model of the type for which this operation was defined
    *
    * @param instance The Predictor instance that we will use to make the predictions
    * @param predictParameters The parameters for the prediction
    * @return A DataSet with the model representation as its only element
    */
  def getModel(instance: Instance, predictParameters: ParameterMap): DataSet[Model]

  /** Calculates the prediction for a single element given the model of the [[Predictor]].
    *
    * @param value The unlabeled example on which we make the prediction
    * @param model The model representation of the prediction algorithm
    * @return A label for the provided example of type [[Prediction]]
    */
  def predict(value: Testing, model: Model):
    Prediction
}

/** Type class for the evaluate operation of [[Predictor]]. This evaluate operation works on
  * DataSets.
  *
  * It takes a [[DataSet]] of some type. For each element of this [[DataSet]] the evaluate method
  * computes the prediction value and returns a tuple of true label value and prediction
```

```
value.
  *
  * @tparam Instance The concrete type of the Predictor instance that we will use to make the
  *                  predictions
  * @tparam Testing The type of the example that we will use to make the predictions (input)
  * @tparam Prediction The type of the label that the prediction operation will produce (output)
  *
  */
trait EvaluateDataSetOperation[Instance, Testing, Prediction] extends Serializable{
  def evaluateDataSet(
      instance: Instance,
      evaluateParameters: ParameterMap,
      testing: DataSet[Testing])
    : DataSet[(Prediction, Prediction)]
}
```

4．Transformer

首先要明确这里的转换操作不同于 DataStream API 和 DataSet API 中的转换操作，前者是考虑数据集总体特征的情况下做出的转换操作，即在模型训练之后的转换操作，如数据预处理 StandardScaler 是在计算整个数据集的均值和方差后再对每一个元素进行转换操作，这也是实时机器学习和批处理机器学习的本质区别之一；而后者仅仅是从数据本身的角度做出的转换操作，如对单个元素的 map、flatMap 操作。

Transformer 继承了前面的 FlinkML 架构思路：首先根据数据集和单个元素分别定义 Transformer 方法原型，在其伴生对象中实现默认的、对数据集的 transform 方法。

此外，由于 Predictor 在一个 Pipeline 中不可能有多个，Transformer 需要定义两个 Transformer 间的 chain 操作，以及 Transformer 和 Predictor 的 chain 操作。

在熟悉了 Estimator 和 Predictor 架构的"套路"后，对 Transformer 的代码分析工作留给读者完成。完整的 Transformer 代码如下：

```
/** Transformer trait for Flink's pipeline operators.
```

```
 *
 * A Transformer transforms a [[DataSet]] of an input type into a [[DataSet]] of an output
type.
 * Furthermore, a [[Transformer]] is also an [[Estimator]], because some transformations
depend
 * on the training data. In order to do that the implementing class has to provide a
 * [[TransformDataSetOperation]] and [[FitOperation]] implementation. The Scala
compiler finds
 * these implicit values if it is put in the scope of the companion object of the
implementing
 * class.
 *
 * [[Transformer]] can be chained with other [[Transformer]] and [[Predictor]] to create
 * pipelines. These pipelines can consist of an arbitrary number of [[Transformer]] and
at most
 * one trailing [[Predictor]].
 *
 * The pipeline mechanism has been inspired by scikit-learn
 *
 * @tparam Self
 */
trait Transformer[Self <: Transformer[Self]]
  extends Estimator[Self]
  with WithParameters
  with Serializable {
  that: Self =>

  /** Transform operation which transforms an input [[DataSet]] of type I into an output
[[DataSet]]
    * of type O. The actual transform operation is implemented within the
    * [[TransformDataSetOperation]].
    *
    * @param input Input [[DataSet]] of type I
    * @param transformParameters Additional parameters for the
[[TransformDataSetOperation]]
    * @param transformOperation [[TransformDataSetOperation]] which encapsulates
    * the algorithm's logic
    * @tparam Input Input data type
    * @tparam Output Output data type
    * @return
    */
  def transform[Input, Output](
      input: DataSet[Input],
```

```
    transformParameters: ParameterMap = ParameterMap.Empty)
   (implicit transformOperation: TransformDataSetOperation[Self, Input, Output])
  : DataSet[Output] = {
  FlinkMLTools.registerFlinkMLTypes(input.getExecutionEnvironment)
  transformOperation.transformDataSet(that, transformParameters, input)
}

/** Chains two [[Transformer]] to form a [[ChainedTransformer]].
  *
  * @param transformer Right side transformer of the resulting pipeline
  * @tparam T Type of the [[Transformer]]
  * @return
  */
def chainTransformer[T <: Transformer[T]](transformer: T): ChainedTransformer[Self, T] = {
  ChainedTransformer(this, transformer)
}

/** Chains a [[Transformer]] with a [[Predictor]] to form a [[ChainedPredictor]].
  *
  * @param predictor Trailing [[Predictor]] of the resulting pipeline
  * @tparam P Type of the [[Predictor]]
  * @return
  */
def chainPredictor[P <: Predictor[P]](predictor: P): ChainedPredictor[Self, P] = {
  ChainedPredictor(this, predictor)
}
}

object Transformer{
 implicit def defaultTransformDataSetOperation[
    Instance <: Estimator[Instance],
    Model,
    Input,
    Output](
    implicit transformOperation: TransformOperation[Instance, Model, Input, Output],
    outputTypeInformation: TypeInformation[Output],
    outputClassTag: ClassTag[Output])
  : TransformDataSetOperation[Instance, Input, Output] = {
   new TransformDataSetOperation[Instance, Input, Output] {
    override def transformDataSet(
        instance: Instance,
        transformParameters: ParameterMap,
```

```
      input: DataSet[Input])
    : DataSet[Output] = {
    val resultingParameters = instance.parameters ++ transformParameters
    val model = transformOperation.getModel(instance, resultingParameters)

    input.mapWithBcVariable(model){
      (element, model) => transformOperation.transform(element, model)
    }
   }
  }
 }
}

/** Type class for a transform operation of [[Transformer]]. This works on [[DataSet]]
of elements.
 *
 * The [[TransformDataSetOperation]] contains a self type parameter so that the Scala compiler
 * looks into the companion object of this class to find implicit values.
 *
 * @tparam Instance Type of the [[Transformer]] for which the
[[TransformDataSetOperation]] is
 *                 defined
 * @tparam Input Input data type
 * @tparam Output Output data type
 */
trait TransformDataSetOperation[Instance, Input, Output] extends Serializable{
  def transformDataSet(
      instance: Instance,
      transformParameters: ParameterMap,
      input: DataSet[Input])
    : DataSet[Output]
}

/** Type class for a transform operation which works on a single element and the corresponding model
 * of the [[Transformer]].
 *
 * @tparam Instance
 * @tparam Model
 * @tparam Input
 * @tparam Output
 */
```

```
trait TransformOperation[Instance, Model, Input, Output] extends Serializable{

 /** Retrieves the model of the [[Transformer]] for which this operation has been defined.
  *
  * @param instance
  * @param transformParameters
  * @return
  */
 def getModel(instance: Instance, transformParameters: ParameterMap): DataSet[Model]

 /** Transforms a single element with respect to the model associated with the respective
  * [[Transformer]]
  *
  * @param element
  * @param model
  * @return
  */
 def transform(element: Input, model: Model): Output
}
```

5．ChainedTransformer

用 L 和 R 表示被 chain 两端的对象。ChainedTransformer 用于 chain 多个 Transformer，形成一个 Transformer 流水线，因此 ChainedTransformer 也是 Transformer 类型；此外，ChainedTransformer 并未定义自己的专有参数，仅定位于管道连接功能，其定义如下：

```
case class ChainedTransformer[L <: Transformer[L], R <: Transformer[R]](left: L, right: R)
  extends Transformer[ChainedTransformer[L, R]] {
}
```

Transformer 有两种方法：fit 和 transform，因此 ChainedTransformer 需要提供对应的 chain 实现。以 transform 为例，ChainedTransformer 先将 Dataset 类型的数据集输入左侧的 Transformer，并将中间结果输入到右侧 Transformer 输出 chain 的两个 Transformer 的转换操作结果，代码如下：

```
implicit def chainedTransformOperation[
    L <: Transformer[L],
    R <: Transformer[R],
```

```
    I,
    T,
    O](implicit
    transformOpLeft: TransformDataSetOperation[L, I, T],
    transformOpRight: TransformDataSetOperation[R, T, O])
  : TransformDataSetOperation[ChainedTransformer[L,R], I, O] = {

  new TransformDataSetOperation[ChainedTransformer[L, R], I, O] {
    override def transformDataSet(
        chain: ChainedTransformer[L, R],
        transformParameters: ParameterMap,
        input: DataSet[I]): DataSet[O] = {
      val intermediateResult = transformOpLeft.transformDataSet(
        chain.left,
        transformParameters,
        input)
      transformOpRight.transformDataSet(chain.right, transformParameters,
intermediateResult)
    }
  }
}
```

完整的 ChainedTransformer 代码如下:

```
/** [[Transformer]] which represents the chaining of two [[Transformer]].
 *
 * A [[ChainedTransformer]] can be treated as regular [[Transformer]]. Upon calling the fit or
 * transform operation, the data is piped through all [[Transformer]] of the pipeline.
 *
 * The pipeline mechanism has been inspired by scikit-learn
 *
 * @param left Left [[Transformer]] of the pipeline
 * @param right Right [[Transformer]] of the pipeline
 * @tparam L Type of the left [[Transformer]]
 * @tparam R Type of the right [[Transformer]]
 */
case class ChainedTransformer[L <: Transformer[L], R <: Transformer[R]](left: L, right: R)
  extends Transformer[ChainedTransformer[L, R]] {
}

object ChainedTransformer{
```

```scala
/** [[TransformDataSetOperation]] implementation for [[ChainedTransformer]].
  *
  * First the transform operation of the left [[Transformer]] is called with the input
data. This
  * generates intermediate data which is fed to the right [[Transformer]]'s transform
operation.
  *
  * @param transformOpLeft [[TransformDataSetOperation]] for the left [[Transformer]]
  * @param transformOpRight [[TransformDataSetOperation]] for the right [[Transformer]]
  * @tparam L Type of the left [[Transformer]]
  * @tparam R Type of the right [[Transformer]]
  * @tparam I Type of the input data
  * @tparam T Type of the intermediate output data
  * @tparam O Type of the output data
  * @return
  */
implicit def chainedTransformOperation[
    L <: Transformer[L],
    R <: Transformer[R],
    I,
    T,
    O](implicit
    transformOpLeft: TransformDataSetOperation[L, I, T],
    transformOpRight: TransformDataSetOperation[R, T, O])
  : TransformDataSetOperation[ChainedTransformer[L,R], I, O] = {

  new TransformDataSetOperation[ChainedTransformer[L, R], I, O] {
    override def transformDataSet(
        chain: ChainedTransformer[L, R],
        transformParameters: ParameterMap,
        input: DataSet[I]): DataSet[O] = {
      val intermediateResult = transformOpLeft.transformDataSet(
        chain.left,
        transformParameters,
        input)
      transformOpRight.transformDataSet(chain.right, transformParameters,
intermediateResult)
    }
  }
}

/** [[FitOperation]] implementation for [[ChainedTransformer]].
```

```
  *
  * First the fit operation of the left [[Transformer]] is called with the input data. Then
  * the data is transformed by this [[Transformer]] and the given to the fit operation of the
  * right [[Transformer]].
  *
  * @param leftFitOperation [[FitOperation]] for the left [[Transformer]]
  * @param leftTransformOperation [[TransformDataSetOperation]] for the left [[Transformer]]
  * @param rightFitOperation [[FitOperation]] for the right [[Transformer]]
  * @tparam L Type of the left [[Transformer]]
  * @tparam R Type of the right [[Transformer]]
  * @tparam I Type of the input data
  * @tparam T Type of the intermediate output data
  * @return
  */
  implicit def chainedFitOperation[L <: Transformer[L], R <: Transformer[R], I, T](implicit
      leftFitOperation: FitOperation[L, I],
      leftTransformOperation: TransformDataSetOperation[L, I, T],
      rightFitOperation: FitOperation[R, T]): FitOperation[ChainedTransformer[L, R], I] = {
    new FitOperation[ChainedTransformer[L, R], I] {
      override def fit(
          instance: ChainedTransformer[L, R],
          fitParameters: ParameterMap,
          input: DataSet[I]): Unit = {
        instance.left.fit(input, fitParameters)
        val intermediateResult = instance.left.transform(input, fitParameters)
        instance.right.fit(intermediateResult, fitParameters)
      }
    }
  }
}
```

6．ChainedPredictor

类似地，ChainedPredictor 也是继承 Predictor，并在其伴生对象中实现了 fit、predict、evaluate 的 chain 操作，对它的代码分析工作留给读者完成。完整的 ChainedPredictor 代码如下：

```scala
case class ChainedPredictor[T <: Transformer[T], P <: Predictor[P]](transformer: T, predictor: P)
  extends Predictor[ChainedPredictor[T, P]]{}

object ChainedPredictor{

  /** [[PredictDataSetOperation]] for the [[ChainedPredictor]].
    *
    * The [[PredictDataSetOperation]] requires the [[TransformDataSetOperation]] of the preceding
    * [[Transformer]] and the [[PredictDataSetOperation]] of the trailing [[Predictor]]. Upon
    * calling predict, the testing data is first transformed by the preceding [[Transformer]] and
    * the result is then used to calculate the prediction via the trailing [[Predictor]].
    *
    * @param transformOperation [[TransformDataSetOperation]] for the preceding [[Transformer]]
    * @param predictOperation [[PredictDataSetOperation]] for the trailing [[Predictor]]
    * @tparam T Type of the preceding [[Transformer]]
    * @tparam P Type of the trailing [[Predictor]]
    * @tparam Testing Type of the testing data
    * @tparam Intermediate Type of the intermediate data produced by the preceding [[Transformer]]
    * @tparam Prediction Type of the predicted data generated by the trailing [[Predictor]]
    * @return
    */
  implicit def chainedPredictOperation[
      T <: Transformer[T],
      P <: Predictor[P],
      Testing,
      Intermediate,
      Prediction](
      implicit transformOperation: TransformDataSetOperation[T, Testing, Intermediate],
      predictOperation: PredictDataSetOperation[P, Intermediate, Prediction])
    : PredictDataSetOperation[ChainedPredictor[T, P], Testing, Prediction] = {

    new PredictDataSetOperation[ChainedPredictor[T, P], Testing, Prediction] {
      override def predictDataSet(
          instance: ChainedPredictor[T, P],
          predictParameters: ParameterMap,
          input: DataSet[Testing])
        : DataSet[Prediction] = {
```

```scala
      val testing = instance.transformer.transform(input, predictParameters)
      instance.predictor.predict(testing, predictParameters)
    }
  }
}

/** [[FitOperation]] for the [[ChainedPredictor]].
  *
  * The [[FitOperation]] requires the [[FitOperation]] and the
[[TransformDataSetOperation]] of
  * the preceding [[Transformer]] as well as the [[FitOperation]] of the trailing
[[Predictor]].
  * Upon calling fit, the preceding [[Transformer]] is first fitted to the training data.
  * The training data is then transformed by the fitted [[Transformer]]. The transformed data
  * is then used to fit the [[Predictor]].
  *
  * @param fitOperation [[FitOperation]] of the preceding [[Transformer]]
  * @param transformOperation [[TransformDataSetOperation]] of the preceding
[[Transformer]]
  * @param predictorFitOperation [[PredictDataSetOperation]] of the trailing
[[Predictor]]
  * @tparam L Type of the preceding [[Transformer]]
  * @tparam R Type of the trailing [[Predictor]]
  * @tparam I Type of the training data
  * @tparam T Type of the intermediate data
  * @return
  */
implicit def chainedFitOperation[L <: Transformer[L], R <: Predictor[R], I, T](implicit
    fitOperation: FitOperation[L, I],
    transformOperation: TransformDataSetOperation[L, I, T],
    predictorFitOperation: FitOperation[R, T]): FitOperation[ChainedPredictor[L, R], I]
= {
  new FitOperation[ChainedPredictor[L, R], I] {
    override def fit(
        instance: ChainedPredictor[L, R],
        fitParameters: ParameterMap,
        input: DataSet[I])
      : Unit = {
      instance.transformer.fit(input, fitParameters)
      val intermediateResult = instance.transformer.transform(input, fitParameters)
      instance.predictor.fit(intermediateResult, fitParameters)
```

```
    }
  }
}

implicit def chainedEvaluationOperation[
    T <: Transformer[T],
    P <: Predictor[P],
    Testing,
    Intermediate,
    PredictionValue](
    implicit transformOperation: TransformDataSetOperation[T, Testing, Intermediate],
    evaluateOperation: EvaluateDataSetOperation[P, Intermediate, PredictionValue],
    testingTypeInformation: TypeInformation[Testing],
    predictionValueTypeInformation: TypeInformation[PredictionValue])
  : EvaluateDataSetOperation[ChainedPredictor[T, P], Testing, PredictionValue] = {
  new EvaluateDataSetOperation[ChainedPredictor[T, P], Testing, PredictionValue] {
    override def evaluateDataSet(
        instance: ChainedPredictor[T, P],
        evaluateParameters: ParameterMap,
        testing: DataSet[Testing])
      : DataSet[(PredictionValue, PredictionValue)] = {
      val intermediate = instance.transformer.transform(testing, evaluateParameters)
      instance.predictor.evaluate(intermediate, evaluateParameters)
    }
  }
}
```

5.3 深入分析多项式曲线拟合

5.3.1 数值计算的底层框架

机器学习的核心是数值计算，而数值计算的核心是向量和矩阵的运算，因此机器学习引擎都有自己对应的底层实现，FlinkML 使用 Breeze 作为其底层计算核心。Breeze 的特点是执行速度快，结合并行 Collection 及 Apache Akka Actor 模型可以构建可伸缩的分布式计算系统，成为 Scala 语言在大数据上应用的利器。

Breeze 是以 Scala 语言编写的，用到了 Scala 语言的类型系统和隐式转化，优化技巧较高，实现了代数运算（如环、域、复数、向量、矩阵的运算），以及 Collection、优化（optimize）、统计、常用算法等基础数学库。此外，Breeze-Viz 可以提供基于 Scala 语言的绘图功能，这是机器学习中重要的可视化工具。

FlinkML 的 Breeze 封装提供了 FlinkML 的（dense 或 sparse）向量矩阵和 Breeze 的向量矩阵互相转换的方法，其中 asBreeze 方法将向量和矩阵转换成 Breeze 的形式以便借助 Breeze 实现高效的数值计算，而 fromBreeze 则可以将运算结果转换成 FlinkML 的内部对象，以便序列化、加载或在内部接口流转。

完整的 Breeze 代码如下：

```
object Breeze {

 implicit class Matrix2BreezeConverter(matrix: Matrix) {
  def asBreeze: BreezeMatrix[Double] = {
   matrix match {
    case dense: DenseMatrix =>
     new BreezeDenseMatrix[Double](
      dense.numRows,
      dense.numCols,
      dense.data)

    case sparse: SparseMatrix =>
     new BreezeCSCMatrix[Double](
      sparse.data,
      sparse.numRows,
      sparse.numCols,
      sparse.colPtrs,
      sparse.rowIndices
     )
   }
  }
 }

 implicit class Breeze2MatrixConverter(matrix: BreezeMatrix[Double]) {
  def fromBreeze: Matrix = {
   matrix match {
    case dense: BreezeDenseMatrix[Double] =>
```

```
          new DenseMatrix(dense.rows, dense.cols, dense.data)

      case sparse: BreezeCSCMatrix[Double] =>
        new SparseMatrix(sparse.rows, sparse.cols, sparse.rowIndices, sparse.colPtrs,
sparse.data)
    }
  }
}

implicit class BreezeArrayConverter[T](array: Array[T]) {
  def asBreeze: BreezeDenseVector[T] = {
    new BreezeDenseVector[T](array)
  }
}

implicit class Breeze2VectorConverter(vector: BreezeVector[Double]) {
  def fromBreeze[T <: Vector: BreezeVectorConverter]: T = {
    val converter = implicitly[BreezeVectorConverter[T]]
    converter.convert(vector)
  }
}

implicit class Vector2BreezeConverter(vector: Vector) {
  def asBreeze: BreezeVector[Double] = {
    vector match {
      case dense: DenseVector =>
        new breeze.linalg.DenseVector(dense.data)

      case sparse: SparseVector =>
        new BreezeSparseVector(sparse.indices, sparse.data, sparse.size)
    }
  }
}
```

5.3.2 向量

向量是机器学习数值计算的主要数据格式，FlinkML 有自己的内部封装形式。

对于向量，FlinkML 定义了长度、某个位置元素的操作、内积与外积、拷贝及从 Breeze 向量转化成 FlinkML 向量对象的方法：

```
trait Vector extends Serializable {
  def size: Int
  def apply(index: Int): Double
  def update(index: Int, value: Double): Unit
  def copy: Vector
  def dot(other: Vector): Double
  def outer(other: Vector): Matrix
  def magnitude: Double

  def equalsVector(vector: Vector): Boolean = {
    if(size == vector.size) {
      (0 until size) forall { idx =>
        this(idx) == vector(idx)
      }
    } else {
      false
    }
  }
}

object Vector{
  implicit val vectorConverter = new BreezeVectorConverter[Vector] {
    override def convert(vector: BreezeVector[Double]): Vector = {
      vector match {
        case dense: BreezeDenseVector[Double] => new DenseVector(dense.data)
        case sparse: BreezeSparseVector[Double] =>
          new SparseVector(
            sparse.length,
            sparse.index.take(sparse.used),
            sparse.data.take(sparse.used))
      }
    }
  }
}
```

并在向量的特质基础上实现了 dense 形式和 sparse 形式，dense 形式向量的代码如下：

```
/**
 * Dense vector implementation of [[Vector]]. The data is represented in a continuous
```

```
array of
 * doubles.
 *
 * @param data Array of doubles to store the vector elements
 */
case class DenseVector(data: Array[Double]) extends Vector with Serializable {

  /**
   * Number of elements in a vector
   * @return the number of the elements in the vector
   */
  override def size: Int = {
    data.length
  }

  /**
   * Element wise access function
   *
   * @param index index of the accessed element
   * @return element at the given index
   */
  override def apply(index: Int): Double = {
    require(0 <= index && index < data.length, index + " not in [0, " + data.length + ")")
    data(index)
  }

  override def toString: String = {
    s"DenseVector(${data.mkString(", ")})"
  }

  override def equals(obj: Any): Boolean = {
    obj match {
      case dense: DenseVector => data.length == dense.data.length &&
data.sameElements(dense.data)
      case _ => false
    }
  }

  override def hashCode: Int = {
    java.util.Arrays.hashCode(data)
  }

  /**
```

```
 * Copies the vector instance
 *
 * @return Copy of the vector instance
 */
override def copy: DenseVector = {
  DenseVector(data.clone())
}

/** Updates the element at the given index with the provided value
  *
  * @param index Index whose value is updated.
  * @param value The value used to update the index.
  */
override def update(index: Int, value: Double): Unit = {
  require(0 <= index && index < data.length, index + " not in [0, " + data.length + ")")

  data(index) = value
}

/** Returns the dot product of the recipient and the argument
  *
  * @param other a Vector
  * @return a scalar double of dot product
  */
override def dot(other: Vector): Double = {
  require(size == other.size, "The size of vector must be equal.")

  other match {
    case SparseVector(_, otherIndices, otherData) =>
      otherIndices.zipWithIndex.map {
        case (idx, sparseIdx) => data(idx) * otherData(sparseIdx)
      }.sum
    case _ => (0 until size).map(i => data(i) * other(i)).sum
  }
}

/** Returns the outer product (a.k.a. Kronecker product) of `this`
  * with `other`. The result will given in [[org.apache.flink.ml.math.SparseMatrix]]
  * representation if `other` is sparse and as [[org.apache.flink.ml.math.DenseMatrix]]
  * otherwise.
  *
  * @param other a Vector
  * @return the [[org.apache.flink.ml.math.Matrix]] which equals the outer product of
```

```scala
 `this`
   *        with `other.`
   */
  override def outer(other: Vector): Matrix = {
    val numRows = size
    val numCols = other.size

    other match {
      case sv: SparseVector =>
        val entries = for {
          i <- 0 until numRows
          (j, k) <- sv.indices.zipWithIndex
          value = this(i) * sv.data(k)
          if value != 0
        } yield (i, j, value)

        SparseMatrix.fromCOO(numRows, numCols, entries)
      case _ =>
        val values = for {
          i <- 0 until numRows
          j <- 0 until numCols
        } yield this(i) * other(j)

        DenseMatrix(numRows, numCols, values.toArray)
    }
  }

  /** Magnitude of a vector
    *
    * @return The length of the vector
    */
  override def magnitude: Double = {
    math.sqrt(data.map(x => x * x).sum)
  }

  /** Convert to a [[SparseVector]]
    *
    * @return Creates a SparseVector from the DenseVector
    */
  def toSparseVector: SparseVector = {
    val nonZero = (0 until size).zip(data).filter(_._2 != 0)

    SparseVector.fromCOO(size, nonZero)
```

```scala
  }
}

object DenseVector {

  def apply(values: Double*): DenseVector = {
    new DenseVector(values.toArray)
  }

  def apply(values: Array[Int]): DenseVector = {
    new DenseVector(values.map(_.toDouble))
  }

  def zeros(size: Int): DenseVector = {
    init(size, 0.0)
  }

  def eye(size: Int): DenseVector = {
    init(size, 1.0)
  }

  def init(size: Int, value: Double): DenseVector = {
    new DenseVector(Array.fill(size)(value))
  }

  /** BreezeVectorConverter implementation for [[org.apache.flink.ml.math.DenseVector]]
    *
    * This allows to convert Breeze vectors into [[DenseVector]].
    */
  implicit val denseVectorConverter = new BreezeVectorConverter[DenseVector] {
    override def convert(vector: BreezeVector[Double]): DenseVector = {
      vector match {
        case dense: BreezeDenseVector[Double] => new DenseVector(dense.data)
        case sparse: BreezeSparseVector[Double] => new DenseVector(sparse.toDenseVector.data)
      }
    }
  }
}
```

5.3.3 数据预处理

通常，假定数据呈现正态分布的形式对于大多数机器学习任务来说是比较安全的，因此正态分布变换是数据预处理的一个重要方法。

StandardScaler 将数据集转化成指定均值和方差的正态分布形式，这涉及以下两个步骤。

（1）计算输入数据集的均值（Mean，数学上用 μ 表示）和方差（Std，数学上用 σ 表示）。显然，这个计算指的是 Transformer 的 fit 方法。这里使用 Youngs & Cramer 算法计算均值和方差，实现代码如下所示：

```scala
private def extractFeatureMetrics[T <: Vector](dataSet: DataSet[T])
: DataSet[(linalg.Vector[Double], linalg.Vector[Double])] = {
  val metrics = dataSet.map{
    v => (1.0, v.asBreeze, linalg.Vector.zeros[Double](v.size))
  }.reduce{
    (metrics1, metrics2) => {
      /* We use formula 1.5b of the cited technical report for the combination of partial
       * sum of squares. According to 1.5b:
       * val temp1 : m/n(m+n)
       * val temp2 : n/m
       */
      val temp1 = metrics1._1 / (metrics2._1 * (metrics1._1 + metrics2._1))
      val temp2 = metrics2._1 / metrics1._1
      val tempVector = (metrics1._2 * temp2) - metrics2._2
      val tempS = (metrics1._3 + metrics2._3) + (tempVector :* tempVector) * temp1

      (metrics1._1 + metrics2._1, metrics1._2 + metrics2._2, tempS)
    }
  }.map{
    metric => {
      val varianceVector = sqrt(metric._3 / metric._1)

      for (i <- 0 until varianceVector.size) {
        if (varianceVector(i) == 0.0) {
          varianceVector.update(i, 1.0)
        }
      }
      (metric._2 / metric._1, varianceVector)
```

```
    }
  }
  metrics
}
```

（2）对每个元素进行变换，以达到整个数据集的均值与方差和设定值一致，这个变换是 Transformer 的 transform 方法。记 μ_0 和 σ_0 为原数据集的均值和方差，μ 和 σ 为均值和方差设定值，其原理用公式表示为

$$X \sim N(\mu_0, \sigma_0) \Rightarrow Y = (X - \mu_0 + \mu) \times \sigma/\sigma_0 \sim N(\mu, \sigma)$$

按照上面的方法转换操作，即 scale 方法定义如下：

```
def scale[V <: Vector: BreezeVectorConverter](
  vector: V,
  model: (linalg.Vector[Double], linalg.Vector[Double]))
: V = {
  val (broadcastMean, broadcastStd) = model
  var myVector = vector.asBreeze
  myVector -= broadcastMean
  myVector :/= broadcastStd
  myVector = (myVector :* std) + mean
  myVector.fromBreeze
}
```

整个实现代码根据输入数据格式实现多个 fit 和 transform 版本，主要代码如下：

```
object StandardScaler {
 /** Trains the [[org.apache.flink.ml.preprocessing.StandardScaler]] by learning the mean and
   * standard deviation of the training data. These values are used inthe transform step
   * to transform the given input data.
   *
   * @tparam T Input data type which is a subtype of [[Vector]]
   * @return
   */
  implicit def fitVectorStandardScaler[T <: Vector] = new FitOperation[StandardScaler, T] {
    override def fit(instance: StandardScaler, fitParameters: ParameterMap, input: DataSet[T]):
```

```scala
      : Unit = {
    val metrics = extractFeatureMetrics(input)

    instance.metricsOption = Some(metrics)
  }
}

/** Trains the [[StandardScaler]] by learning the mean and standard deviation of the training
  * data which is of type [[LabeledVector]]. The mean and standard deviation are used to
  * transform the given input data.
  *
  */
implicit val fitLabeledVectorStandardScaler = {
  new FitOperation[StandardScaler, LabeledVector] {
    override def fit(
        instance: StandardScaler,
        fitParameters: ParameterMap,
        input: DataSet[LabeledVector])
      : Unit = {
      val vectorDS = input.map(_.vector)
      val metrics = extractFeatureMetrics(vectorDS)

      instance.metricsOption = Some(metrics)
    }
  }
}

/** Trains the [[StandardScaler]] by learning the mean and standard deviation of the training
  * data which is of type ([[Vector]], Double). The mean and standard deviation are used to
  * transform the given input data.
  *
  */
implicit def fitLabelVectorTupleStandardScaler
[T <: Vector: BreezeVectorConverter: TypeInformation: ClassTag] = {
  new FitOperation[StandardScaler, (T, Double)] {
    override def fit(
        instance: StandardScaler,
        fitParameters: ParameterMap,
        input: DataSet[(T, Double)])
```

```scala
    : Unit = {
      val vectorDS = input.map(_._1)
      val metrics = extractFeatureMetrics(vectorDS)

      instance.metricsOption = Some(metrics)
    }
  }
}

/** Calculates in one pass over the data the features' mean and standard deviation.
  * For the calculation of the Standard deviation with one pass over the data,
  * the Youngs & Cramer algorithm was used:
  * [[http://www.cs.yale.edu/publications/techreports/tr222.pdf]]
  *
  *
  * @param dataSet The data set for which we want to calculate mean and variance
  * @return DataSet containing a single tuple of two vectors (meanVector, stdVector).
  *         The first vector represents the mean vector and the second is the standard
  *         deviation vector.
  */
private def extractFeatureMetrics[T <: Vector](dataSet: DataSet[T])
: DataSet[(linalg.Vector[Double], linalg.Vector[Double])] = {
  val metrics = dataSet.map{
    v => (1.0, v.asBreeze, linalg.Vector.zeros[Double](v.size))
  }.reduce{
    (metrics1, metrics2) => {
      /* We use formula 1.5b of the cited technical report for the combination of partial
        * sum of squares. According to 1.5b:
        * val temp1 : m/n(m+n)
        * val temp2 : n/m
        */
      val temp1 = metrics1._1 / (metrics2._1 * (metrics1._1 + metrics2._1))
      val temp2 = metrics2._1 / metrics1._1
      val tempVector = (metrics1._2 * temp2) - metrics2._2
      val tempS = (metrics1._3 + metrics2._3) + (tempVector :* tempVector) * temp1

      (metrics1._1 + metrics2._1, metrics1._2 + metrics2._2, tempS)
    }
  }.map{
    metric => {
      val varianceVector = sqrt(metric._3 / metric._1)

      for (i <- 0 until varianceVector.size) {
```

```
          if (varianceVector(i) == 0.0) {
            varianceVector.update(i, 1.0)
          }
        }
        (metric._2 / metric._1, varianceVector)
      }
    }
  metrics
}

/** Base class for StandardScaler's [[TransformOperation]]. This class has to be extended for
  * all types which are supported by [[StandardScaler]]'s transform operation.
  *
  * @tparam T
  */
abstract class StandardScalerTransformOperation[T: TypeInformation: ClassTag]
  extends TransformOperation[
    StandardScaler,
    (linalg.Vector[Double], linalg.Vector[Double]),
    T,
    T] {

  var mean: Double = _
  var std: Double = _

  override def getModel(
    instance: StandardScaler,
    transformParameters: ParameterMap)
  : DataSet[(linalg.Vector[Double], linalg.Vector[Double])] = {
    mean = transformParameters(Mean)
    std = transformParameters(Std)

    instance.metricsOption match {
      case Some(metrics) => metrics
      case None =>
        throw new RuntimeException("The StandardScaler has not been fitted to the data. " +
"This is necessary to estimate the mean and standard deviation of the data.")
    }
  }

  def scale[V <: Vector: BreezeVectorConverter](
```

```scala
    vector: V,
    model: (linalg.Vector[Double], linalg.Vector[Double]))
  : V = {
    val (broadcastMean, broadcastStd) = model
    var myVector = vector.asBreeze
    myVector -= broadcastMean
    myVector :/= broadcastStd
    myVector = (myVector :* std) + mean
    myVector.fromBreeze
  }
}

/** [[TransformOperation]] to transform [[Vector]] types
  *
  * @tparam T
  * @return
  */
implicit def transformVectors[T <: Vector: BreezeVectorConverter: TypeInformation:
ClassTag] = {
  new StandardScalerTransformOperation[T]() {
    override def transform(
        vector: T,
        model: (linalg.Vector[Double], linalg.Vector[Double]))
      : T = {
      scale(vector, model)
    }
  }
}

/** [[TransformOperation]] to transform tuples of type ([[Vector]], [[Double]]).
  *
  * @tparam T
  * @return
  */
implicit def transformTupleVectorDouble[
    T <: Vector: BreezeVectorConverter: TypeInformation: ClassTag] = {
  new StandardScalerTransformOperation[(T, Double)] {
    override def transform(
        element: (T, Double),
        model: (linalg.Vector[Double], linalg.Vector[Double]))
      : (T, Double) = {
      (scale(element._1, model), element._2)
    }
```

```
    }
  }

  /** [[TransformOperation]] to transform [[LabeledVector]].
    *
    */
  implicit val transformLabeledVector = new
StandardScalerTransformOperation[LabeledVector] {
    override def transform(
      element: LabeledVector,
      model: (linalg.Vector[Double], linalg.Vector[Double]))
      : LabeledVector = {
      val LabeledVector(label, vector) = element

      LabeledVector(label, scale(vector, model))
    }
  }
}
```

5.3.4 特征变换

特征变换的作用是将输入数据转换成高维特征空间的形式，因此每个输入向量的长度发生了变化，且 PolynomialFeatures 不需要 fit 方法。

PolynomialFeatures 的两个关键计算步骤如下。

（1）计算给定向量长度和指数和的组合形式，如在 $n=2$ 的二元多项式拟合时，指定指数和为 2，则对应的项有 (x_1^2, x_1x_2, x_2^2)，因此本步骤计算的结果为向量 $((2,0),(1,1),(0,2))$，输出结果为 List[List[Int]]：

```
private def calculateCombinations(length: Int, value: Int): List[List[Int]] = {
  if(length == 0) {
    List()
  } else if (length == 1) {
    List(List(value))
  } else {
    value to 0 by -1 flatMap {
      v =>
        calculateCombinations(length - 1, value - v) map {
          v::_
```

```
    }
  } toList
 }
}
```

（2）递归输出每个指数和的特征向量，根据指数和从高到低依次生成，对应上面假定的指数和为 2 的情况，输出 (x_1^2, x_1x_2, x_2^2)，最后将每个指数和的向量拼接成一个特征向量：

```
private def calculateCombinedCombinations(degree: Int, vector: Vector): List[Double]
= {
  if(degree == 0) {
    List()
  } else {
    val partialResult = calculateCombinedCombinations(degree - 1, vector)
    val combinations = calculateCombinations(vector.size, degree)
    val result = combinations map {
      combination =>
        combination.zipWithIndex.map{
          case (exp, idx) => math.pow(vector(idx), exp)
        }.fold(1.0)(_ * _)
    }
    result ::: partialResult
  }
}
```

整个实现代码根据输入数据格式实现多个 transform 版本，主要代码如下：

```
object PolynomialFeatures{
  /** The [[PolynomialFeatures]] transformer does not need a fitting phase.
   *
   * @tparam T The fitting works with arbitrary input types
   * @return
   */
  implicit def fitNoOp[T] = {
    new FitOperation[PolynomialFeatures, T]{
      override def fit(
          instance: PolynomialFeatures,
          fitParameters: ParameterMap,
          input: DataSet[T])
        : Unit = {}
    }
```

```scala
  }

  /** [[org.apache.flink.ml.pipeline.TransformDataSetOperation]] to map a [[Vector]] into the
    * polynomial feature space.
    *
    * @tparam T Subclass of [[Vector]]
    * @return
    */
  implicit def transformVectorIntoPolynomialBase[
    T <: Vector : VectorBuilder: TypeInformation: ClassTag
  ] = {
    new TransformDataSetOperation[PolynomialFeatures, T, T] {
      override def transformDataSet(
          instance: PolynomialFeatures,
          transformParameters: ParameterMap,
          input: DataSet[T])
        : DataSet[T] = {
        val resultingParameters = instance.parameters ++ transformParameters

        val degree = resultingParameters(Degree)

        input.map {
          vector => {
            calculatePolynomial(degree, vector)
          }
        }
      }
    }
  }

  /** [[org.apache.flink.ml.pipeline.TransformDataSetOperation]] to map a [[LabeledVector]] into the
    * polynomial feature space
    */
  implicit val transformLabeledVectorIntoPolynomialBase =
    new TransformDataSetOperation[PolynomialFeatures, LabeledVector, LabeledVector] {

      override def transformDataSet(
          instance: PolynomialFeatures,
          transformParameters: ParameterMap,
          input: DataSet[LabeledVector])
        : DataSet[LabeledVector] = {
```

```
    val resultingParameters = instance.parameters ++ transformParameters

    val degree = resultingParameters(Degree)

    input.map {
      labeledVector => {
        val vector = labeledVector.vector
        val label = labeledVector.label

        val transformedVector = calculatePolynomial(degree, vector)

        LabeledVector(label, transformedVector)
      }
    }
  }
}

private def calculatePolynomial[T <: Vector: VectorBuilder](degree: Int, vector: T): T = {
  val builder = implicitly[VectorBuilder[T]]
  builder.build(calculateCombinedCombinations(degree, vector))
}

/** Calculates for a given vector its representation in the polynomial feature space.
  *
  * @param degree Maximum degree of polynomial
  * @param vector Values of the polynomial variables
  * @return List of polynomial values
  */
private def calculateCombinedCombinations(degree: Int, vector: Vector): List[Double] = {
  if(degree == 0) {
    List()
  } else {
    val partialResult = calculateCombinedCombinations(degree - 1, vector)

    val combinations = calculateCombinations(vector.size, degree)

    val result = combinations map {
      combination =>
        combination.zipWithIndex.map{
          case (exp, idx) => math.pow(vector(idx), exp)
```

```
      }.fold(1.0)(_ * _)
    }

    result ::: partialResult
  }


}

/** Calculates all possible combinations of a polynom of degree `value`, whereas the polynom
  * can consist of up to `length` factors. The return value is the list of the exponents of the
  * individual factors
  *
  * @param length maximum number of factors
  * @param value degree of polynomial
  * @return List of lists which contain the exponents of the individual factors
  */
private def calculateCombinations(length: Int, value: Int): List[List[Int]] = {
  if(length == 0) {
    List()
  } else if (length == 1) {
    List(List(value))
  } else {
    value to 0 by -1 flatMap {
      v =>
        calculateCombinations(length - 1, value - v) map {
          v::_
        }
    } toList
  }
}
```

5.3.5 线性拟合

多项式线性拟合并没有使用闭式解，而是采用基于梯度下降的优化方法，采用线性模型的均方误差定义损失函数。由于采用数值解法，迭代次数、步长、收敛阈值、学习速率等参数都需要设定，否则 MultipleLinearRegression 将采用默认值。这些参数即模型的超参，而训练出来的权重值 w 则是模型参数，优化解法相关内容见 5.4 节，这里不再详细分析，以下是使用拟合方法的代码：

```scala
implicit val fitMLR = new FitOperation[MultipleLinearRegression, LabeledVector] {
  override def fit(
    instance: MultipleLinearRegression,
    fitParameters: ParameterMap,
    input: DataSet[LabeledVector])
  : Unit = {
    val map = instance.parameters ++ fitParameters
    // retrieve parameters of the algorithm
    val numberOfIterations = map(Iterations)
    val stepsize = map(Stepsize)
    val convergenceThreshold = map.get(ConvergenceThreshold)
    val learningRateMethod = map.get(LearningRateMethodValue)
    val lossFunction = GenericLossFunction(SquaredLoss, LinearPrediction)
    val optimizer = GradientDescent()
      .setIterations(numberOfIterations)
      .setStepsize(stepsize)
      .setLossFunction(lossFunction)
    convergenceThreshold match {
      case Some(threshold) => optimizer.setConvergenceThreshold(threshold)
      case None =>
    }
    learningRateMethod match {
      case Some(method) => optimizer.setLearningRateMethod(method)
      case None =>
    }
    instance.weightsOption = Some(optimizer.optimize(input, None))
  }
}
```

predict 方法计算权重参数和测试向量的点积得到输出的预测值：

```scala
override def predict(value: T, model: WeightVector): Double = {
  import Breeze._
  val WeightVector(weights, weight0) = model
  val dotProduct = value.asBreeze.dot(weights.asBreeze)
  dotProduct + weight0
}
```

5.4 分类算法

我们先从分类算法要解决的问题开始分析,即用最优超平面理论解决欧式空间中的二分任务。

5.4.1 最优超平面

欧式空间中的二分任务定义为:根据已标注的训练样本集求得一个最优超平面,使得基于此超平面进行分类预测时的泛化性能最优,且此泛化性能基于最小化结构风险(SR,Structural Risk)而非最小化经验风险(ER,Empirical Risk),其中训练样本集定义为

$$D = \{x_1, x_2, ..., x_m\}$$

分别定义 D_+ 和 D_- 为两类样本集,则欧式空间中的二分任务可形象化地描述为:空间内有两类点,欲找到一个最优超平面将两者分开,如图 5-1 所示。

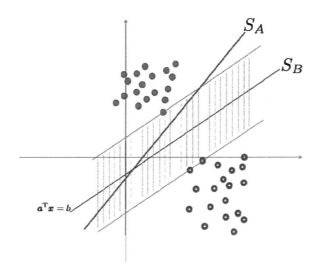

图 5-1 最优超平面分类

有两个超平面 S_A 和 S_B 均能满足要求,怎么比较两者的分类性能差异呢?可以

沿某个超平面平行的方向构建隔离带，使得两类点在隔离带之外，最大隔离带宽度对应的超平面是最优的。同时可以看出，此超平面的方向是决定性因素。如果两类样本点出现的概率相差不多，那么最优超平面应选择最大隔离带中间的位置，如图 5-2 所示。

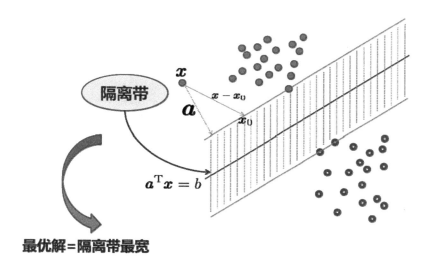

图 5-2　对最优超平面分类的分析

下面对其进行数学描述。

超平面定义为

$$\boldsymbol{a}^{\mathrm{T}}\boldsymbol{x} = b, \boldsymbol{a} \in \mathbb{R}^n \quad \boldsymbol{x} \in \mathbb{R}^n \tag{5.1}$$

点到超平面的距离（$\boldsymbol{x_0}$ 为超平面上任意一点）为

$$\begin{aligned} d_{\boldsymbol{x}} &= \frac{\|\boldsymbol{a}^{\mathrm{T}}(\boldsymbol{x} - \boldsymbol{x_0})\|}{\|\boldsymbol{a}\|} \\ &= \frac{\|\boldsymbol{a}^{\mathrm{T}}\boldsymbol{x} - b\|}{\|\boldsymbol{a}\|} \end{aligned} \tag{5.2}$$

对于任意可行的分类超平面，定义如下两个集合为

$$d_+ = \inf\{d_{\boldsymbol{x}} | \boldsymbol{x} \in D_+\}$$
$$= d_{\boldsymbol{x_1}}$$
$$d_- = \inf\{d_{\boldsymbol{x}} | \boldsymbol{x} \in D_-\}$$
$$= d_{\boldsymbol{x_2}}$$

根据前面的分析，可选择合适的位置参数 $\boldsymbol{b_0}$，使得

$$\boldsymbol{d_+} = \boldsymbol{d_-}$$

即

$$\|\boldsymbol{a}^{\mathrm{T}}\boldsymbol{x_1} - b_0\| = \|\boldsymbol{a}^{\mathrm{T}}\boldsymbol{x_2} - b_0\|$$

此外，对可行超平面，定义缩放变换为

$$(\boldsymbol{a}, b) \mapsto (\lambda \boldsymbol{a}, \lambda b)$$

超平面本身并没有发生变化，因此可以令

$$\|\boldsymbol{a}^{\mathrm{T}}\boldsymbol{x_1} - b_0\| = \|\boldsymbol{a}^{\mathrm{T}}\boldsymbol{x_2} - b_0\| = 1$$

这样，如果定义样本标记空间为

$$\mathcal{Y} = \{+1, -1\}$$

则分类任务的数学形式为

$$\underset{(\boldsymbol{a},b)}{\operatorname{argmax}} \frac{2}{\|\boldsymbol{a}\|}$$
$$\text{s.t.} \quad y_i(\boldsymbol{a}^{\mathrm{T}}\boldsymbol{x} - b) \geqslant 1$$

为了求解方便，可将问题等价为

$$\underset{(\boldsymbol{a},b)}{\operatorname{argmin}} \frac{1}{2}\|\boldsymbol{a}\|^2 \tag{5.3}$$
$$\text{s.t.} \quad y_i(\boldsymbol{a}^{\mathrm{T}}\boldsymbol{x} - b) \geqslant 1$$

其中，使得约束条件中等号成立的点决定了最优分类超平面，称为支持向量，

这也是支持向量机（SVM，Support-Vector Machine）名称的内涵。

表现复杂问题的数学形式往往是简单的，这就是数学之美的体现，牛顿的古典物理学定理如此，支持向量机也如此。凸优化理论正是解决上述条件极值问题的数学武器。

5.4.2 凸优化理论

以**投资组合优化问题**为例：在投资总额固定的情况下，求投资在 n 种资产中最佳的分配方案。可行方案的特点，**一是**总体收益不低于某个期望；**二是**投资风险最小，此问题就是要求条件极值。

1. 数学优化

数学优化的一般形式定义为

$$\begin{aligned}
&\min \quad f_0(\boldsymbol{x}) \\
&\text{s.t.} \quad f_i(\boldsymbol{x}) \leqslant b_i, \quad i = 1, ..., m
\end{aligned} \tag{5.4}$$

其中

$$f_i : \mathbb{R}^n \longrightarrow \mathbb{R}, \quad i = 0, 1, ..., m$$

目标是求得 $\boldsymbol{x}^*$，使得

$$\forall \boldsymbol{z} \quad (f_i(\boldsymbol{x}) \leqslant b_i \quad (i = 1, ..., m)), f_0(\boldsymbol{z}) \geqslant f_0(\boldsymbol{x}^*)$$

定义

$$f_0(\boldsymbol{x})$$

为目标函数。

定义

$$f_i(\boldsymbol{x}), \quad i = 1, ..., m$$

为约束函数。

定义

$$b_i, \quad i=1,\dots,m$$

为约束边界。

定义

$$p^* = f_0(x^*)$$

为最优解。

线性规划的形式为

$$\min \quad c^\mathrm{T} x$$
$$\text{s.t.} \quad Gx \preceq h, \quad Ax = b$$

二次优化的定义为

$$\min \quad \frac{1}{2} x^\mathrm{T} P x + q^\mathrm{T} x + r$$
$$\text{s.t.} \quad Gx \preceq h, \quad Ax = b$$

2. 凸优化

当目标函数与约束函数均为凸函数时，数学优化即为凸优化。

凸函数的定义为

$$\forall \alpha, \beta \geqslant 0, \quad \alpha + \beta = 1$$
$$f(\alpha x + \beta y) \leqslant \alpha f(x) + \beta f(y), \quad f(x): \mathbb{R}^n \to \mathbb{R}$$

如图 5-3 所示。

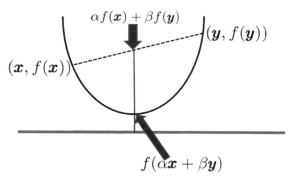

图 5-3 凸函数的定义

凸集的定义为

$$\forall \alpha, \beta \geqslant 0, \quad \alpha + \beta = 1$$
$$\forall \boldsymbol{x}, \boldsymbol{y} \in C \quad \Rightarrow \quad \alpha \boldsymbol{x} + \beta \boldsymbol{y} \in C$$

即集合 C 中任意两个点的线段仍然在集合 C 中。

凸集与凸函数的优良数学特性、基于这些性质导出的可计算性及凸优化问题求解的高效性，使得凸优化成为机器学习领域计算建模的强有力工具之一。例如，凸集的分离与支撑超平面，在理论上解决本章所述的最优超平面解决二分任务是可行的，凸函数的全局分析特性，使得凸优化的数值方法取得累累硕果。以下介绍两个与凸集相关的定理，通过两个定理的描述，读者可以深刻地理解最优超平面分类方法的可行性，定理的证明可参考相关资料。

1）超平面分离定理

假设 C 和 D 是两个不相交的凸集，那么存在非零向量 $\boldsymbol{a}$ 及 b 使得下列两式同时成立。

$$\forall \boldsymbol{x} \in C, \quad \boldsymbol{a}^\mathrm{T} \boldsymbol{x} \leqslant b$$
$$\forall \boldsymbol{y} \in D, \quad \boldsymbol{a}^\mathrm{T} \boldsymbol{y} \geqslant b$$

则定义超平面 $\{\boldsymbol{x} | \boldsymbol{a}^\mathrm{T} \boldsymbol{x} = b\}$ 为集合 C 和 D 的分离超平面。

2）支撑超平面定理

设 $\boldsymbol{x}_0$ 为 n 维空间的集合 C 边界上的一点，满足下列条件

$$\exists \boldsymbol{a} \in \mathbb{R}^n, \quad \text{s.t.}$$
$$\forall \boldsymbol{x} \in C, \boldsymbol{a}^\mathrm{T}\boldsymbol{x} \leqslant \boldsymbol{a}^\mathrm{T}\boldsymbol{x}_0$$

那么称超平面

$$\{\boldsymbol{x} | \boldsymbol{a}^\mathrm{T}\boldsymbol{x} = \boldsymbol{a}^\mathrm{T}\boldsymbol{x}_0\}$$

为集合 C 在 $\boldsymbol{x}_0$ 处的支撑超平面。

从上面两个定理可以分析出，在训练样本集可分的情况下，使用最优超平面方法进行分类是可行的。

3. 对偶

修辞中的对偶是用字数相等、结构相同、意义对称的一对短语或句子来表达两个相对应或相近，或意思相同的修辞方式，在数学理论中也存在相似的理论。在遇到原问题域难解决的问题时，可通过变换的方式，等价出一个对偶问题，以便问题在对偶域易于解决。

拉格朗日对偶函数及 KKT（Karush-Kuhn-Tucker）最优性条件在凸优化里占有重要的地位。将数学优化一般形式中的等式约束和不等式约束拆分成两类，重新定义为

$$\begin{aligned}\min \quad & f_0(\boldsymbol{x})\\ \text{s.t.} \quad & f_i(\boldsymbol{x}) \leqslant b_i, \quad i=1,...,m \\ & h_i(\boldsymbol{x}) = 0, \quad i=1,...,p\end{aligned} \tag{5.5}$$

拉格朗日对偶的基本思想是在目标函数中加入约束条件，得到拉格朗日函数（下称对偶函数）

$$L(\boldsymbol{x}, \boldsymbol{\Lambda}, \boldsymbol{\nu}) = f_0(\boldsymbol{x}) + \sum_{i=1}^{m}\lambda_i f_i(\boldsymbol{x}) + \sum_{i=1}^{p}\nu_i h_i(\boldsymbol{x}) \tag{5.6}$$

其中
$$\Lambda = (\lambda_1, \lambda_2, ..., \lambda_m)^{\mathrm{T}}$$
$$\nu = (\nu_1, \nu_2, ..., \nu_p)^{\mathrm{T}}$$

则数学优化问题等价为

$$p^* = \sup_{(\Lambda, \nu)} \inf_{x \in \mathcal{D}} L(x, \Lambda, \nu) \tag{5.7}$$

其中，定义域为原问题目标函数和约束函数定义域的交集，即

$$\mathcal{D} = \bigcap_{i=0}^{m} \mathrm{dom}(f_i) \bigcap_{i=1}^{m} \mathrm{dom}(h_i)$$

假设目标函数和约束条件函数都是可微函数，令 x^* 和 (Λ^*, ν^*) 为原问题和对偶问题的最优解，则最优解应满足 KKT 条件，

$$\begin{aligned} f_i(x^*) &\leqslant 0, \quad i = 1, ..., m \\ h_i(x^*) &= 0, \quad i = 1, ..., p \\ \lambda_i^* &\geqslant 0, \quad i = 1, ..., m \\ \lambda_i^* f_i(x^*) &= 0, \quad i = 1, ..., m \\ \nabla f_0(x^*) + \sum_{i=1}^{m} \lambda_i^* \nabla f_i(x^*) &+ \sum_{i=1}^{p} \nu_i^* \nabla h_i(x^*) = 0 \end{aligned} \tag{5.8}$$

值得注意的是，上述描述的对偶及 KKT 条件并没有假设目标函数和约束条件为凸函数。

4. 凸优化的数值方法

定义函数 f 为 $\mathbf{R}^n$ 到 $\mathbf{R}$ 的二次可微凸函数，在定义域内的最小值（最优值）存在且唯一，即存在唯一的最优点 x^*，并用 p^* 表示最优值，则最优点满足以下充要条件：

$$\nabla f(x^*) = 0$$

在一般情况下，上述方程无法通过解析的方法求解，但可以通过基于迭代算

法原理的数值方法求解,即在其定义域内寻找点列$x^{(0)},x^{(1)},\ldots,x^{(k)}$,使得以下极限成立:

$$\lim_{k\to\infty} f(x^{(k)}) = p^*$$

且当

$$f(x^{(k)}) - p^* \leqslant \varepsilon, \quad \varepsilon > 0$$

时算法终止,其中ε为设定的容许误差。

为了加快收敛速度,点列是在合理的搜索方向上取得的,即搜索算法表征了相应的数值解法,如牛顿法、梯度下降法等。梯度下降求解过程,如图 5-4 所示。

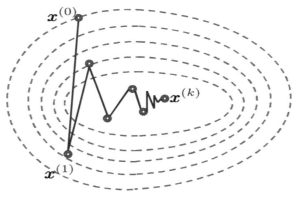

图 5-4 梯度下降求解过程

5.4.3 求解最优超平面

根据拉格朗日对偶及 KKT 条件,下面继续推导(5.3)式所定义问题的数学形式。对偶函数为

$$L(\boldsymbol{a},b,\boldsymbol{\Lambda}) = \frac{1}{2}\|\boldsymbol{a}\|^2 + \sum_{i=1}^{m}\lambda_i(1-y_i(\boldsymbol{a}^{\mathrm{T}}\boldsymbol{x}_i - b))$$
$$= \frac{1}{2}\mathrm{tr}(\boldsymbol{a}^{\mathrm{T}}\boldsymbol{a}) + \sum_{i=1}^{m}\lambda_i(1-y_i(\boldsymbol{a}^{\mathrm{T}}\boldsymbol{x}_i - b))$$
$$= \frac{1}{2}\boldsymbol{a}^{\mathrm{T}}\boldsymbol{a} + \sum_{i=1}^{m}\lambda_i(1-y_i(\boldsymbol{a}^{\mathrm{T}}\boldsymbol{x}_i - b))$$

在原问题定义域求极值，即

$$\frac{\partial}{\partial \boldsymbol{a}}L(\boldsymbol{a},b,\boldsymbol{\Lambda}) = \boldsymbol{a} + \sum_{i=1}^{m}\lambda_i(0-y_i(\boldsymbol{x}_i-0)) = 0$$
$$\frac{\partial}{\partial b}L(\boldsymbol{a},b,\boldsymbol{\Lambda}) = 0 + \sum_{i=1}^{m}\lambda_i(0-y_i(0-1)) = 0$$

可求得

$$\boldsymbol{a} = \sum_{i=1}^{m}\lambda_i y_i \boldsymbol{x}_i$$
$$0 = \sum_{i=1}^{m}\lambda_i y_i$$

代入对偶函数，得

$$L(\boldsymbol{\Lambda}) = L(\boldsymbol{a},b,\boldsymbol{\Lambda})$$
$$= \sum_{i=1}^{m}\lambda_i - \frac{1}{2}\boldsymbol{a}^{\mathrm{T}}\boldsymbol{a}$$
$$= \sum_{i=1}^{m}\lambda_i - \frac{1}{2}\sum_{i=1}^{m}\sum_{j=1}^{m}\lambda_i\lambda_j y_i y_j \boldsymbol{x}_i^{\mathrm{T}}\boldsymbol{x}_j$$

令

$$\boldsymbol{A} = (y_1\boldsymbol{x}_1, y_2\boldsymbol{x}_2, y_i\boldsymbol{x}_i, ..., y_m\boldsymbol{x}_m)$$
$$\boldsymbol{P} = \boldsymbol{A}^{\mathrm{T}}\boldsymbol{A}$$
$$\boldsymbol{y} = (y_1, y_2, ..., y_m)^{\mathrm{T}}$$

则可进一步简化对偶函数的形式为

$$L(\mathbf{\Lambda}) = -\frac{1}{2}\sum_{i=1}^{m}\sum_{j=1}^{m}\lambda_i\lambda_j y_i y_j \boldsymbol{x}_i^{\mathrm{T}}\boldsymbol{x}_j + \sum_{i=1}^{m}\lambda_i$$
$$= \frac{1}{2}\mathbf{\Lambda}^{\mathrm{T}}(\boldsymbol{P})\mathbf{\Lambda} - (1,1,...,1)\mathbf{\Lambda}$$

根据对偶理论，原问题等价于求解上式的极大值，为了分析方便，把上式变成求解最小值的形式为

$$\begin{aligned}\min\quad W(\mathbf{\Lambda}) &= -L(\mathbf{\Lambda}) \\ &= \frac{1}{2}\mathbf{\Lambda}^{\mathrm{T}}(\boldsymbol{P})\mathbf{\Lambda} - \mathbf{\Lambda}^{\mathrm{T}}\mathbf{1}\end{aligned} \tag{5.9}$$

约束条件为

$$\begin{aligned}\mathbf{\Lambda}^{\mathrm{T}}\boldsymbol{y} &= 0 \\ -\mathbf{\Lambda} &\preceq 0 \\ \lambda_i(1-y_if(\boldsymbol{x}_i)) &= 0\end{aligned} \tag{5.10}$$

以上问题可使用凸优化的数值方法求解。由于以上问题的维度是训练样本集的大小，求解难度非常大，在后面我们会研究优化解法。

5.4.4 核方法

最优超平面方法假定训练样本集是可分的，对于不可分的情况，可先对样本集进行非线性映射，在这个高维空间中找到超平面，使得训练样本集可分。

先举一个在二维平面上的例子加以说明。

设有两类训练样本分别均匀地分布在半径为 3 和 5 的圆周上，这两类样本不是线性可分的。但是按照最优超平面思想，选择半径为 4 的圆周是这个分类任务最理想的结果。为了得到高维空间的结果，我们做如下映射：

$$(x, y) \mapsto (x, y, \sqrt{x^2 + y^2})$$

在映射的三维空间中，这两类训练样本是线性可分的，且最优超平面为

$$z = 4$$

图 5-5 中左侧为待分类的数据集，右侧为高维空间映射情况。

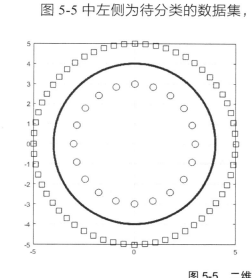
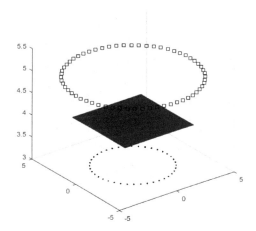

图 5-5　二维单位圆的高维分类

以上例子有数学理论的支持。基于代数拓扑学的分析，对于不可分情形，高维空间的线性超平面是存在的。考虑将训练样本集映射为流形的顶点，在对应的高维空间中，训练集是可分的，但是这种映射的泛化能力很差。

再生核希尔伯特空间理论为此类非线性方法指明了方向。在原样本空间中做如下变换：

$$\boldsymbol{x} \mapsto \boldsymbol{\phi}(\boldsymbol{x}): \quad \mathbb{R}^n \to \mathbb{R}^N$$

代入最优超平面问题，则对偶后的形式没有变化，即

$$L(\boldsymbol{\Lambda}) = \sum_{i=1}^{m} \lambda_i - \frac{1}{2} \sum_{i=1}^{m} \sum_{j=1}^{m} \lambda_i \lambda_j y_i y_j \boldsymbol{\phi}(\boldsymbol{x}_i)^{\mathrm{T}} \boldsymbol{\phi}(\boldsymbol{x}_j) \tag{5.11}$$

其中

$$\boldsymbol{a} = \sum_{i=1}^{m} \lambda_i y_i \phi(\boldsymbol{x}_i)$$

$$0 = \sum_{i=1}^{m} \lambda_i y_i$$

写成矩阵的形式为

$$\min \quad W(\boldsymbol{\Lambda}) = -L(\boldsymbol{\Lambda})$$
$$= \frac{1}{2}\boldsymbol{\Lambda}^{\mathrm{T}}(\boldsymbol{P})\boldsymbol{\Lambda} - \boldsymbol{\Lambda}^{\mathrm{T}}\mathbf{1}$$

约束条件为

$$\boldsymbol{\Lambda}^{\mathrm{T}}\boldsymbol{y} = 0$$
$$-\boldsymbol{\Lambda} \preceq 0$$
$$\lambda_i(1 - y_i f(\phi(\boldsymbol{x}_i))) = 0$$

其中

$$\boldsymbol{A} = (y_1\phi(\boldsymbol{x}_1), y_2\phi(\boldsymbol{x}_2), y_i\phi(\boldsymbol{x}_i), ..., y_m\phi(\boldsymbol{x}_m))$$
$$\boldsymbol{P} = \boldsymbol{A}^{\mathrm{T}}\boldsymbol{A}$$
$$\boldsymbol{y} = (y_1, y_2, ..., y_m)^{\mathrm{T}}$$

(5.11)式的优化需要在映射后的高维特征空间中进行矩阵计算（内积运算）。为了避免维度灾难，我们定义核函数为一个正定对称函数$K(\boldsymbol{u}, \boldsymbol{v})$，使得

$$K(\boldsymbol{x}_i, \boldsymbol{x}_j) = \phi(\boldsymbol{x}_i)^{\mathrm{T}}\phi(\boldsymbol{x}_j) = <\phi(\boldsymbol{x}_i), \phi(\boldsymbol{x}_j)>, \forall i, j < m$$

进一步分析发现，每一个上述核函数都对应一个由函数构成的Hilbert空间，且此函数可以生成对应的Hilbert空间。常用核函数如下。

（1）线性核函数：

$$K(\boldsymbol{x}_i, \boldsymbol{x}_j) = \boldsymbol{x}_i^{\mathrm{T}}\boldsymbol{x}_j$$

（2）多项式核函数：

$$K(\boldsymbol{x}_i, \boldsymbol{x}_j) = (\boldsymbol{x}_i^{\mathrm{T}} \boldsymbol{x}_j)^d$$

（3）高斯（RBF）核函数：

$$K(\boldsymbol{x}_i, \boldsymbol{x}_j) = \exp(-\frac{\|\boldsymbol{x}_i - \boldsymbol{x}_j\|}{\sigma}) \quad \sigma > 0$$

（4）Sigmoid 核函数：

$$K(\boldsymbol{x}_i, \boldsymbol{x}_j) = \tanh(\beta \boldsymbol{x}_i^{\mathrm{T}} \boldsymbol{x}_j + \theta) \quad \beta > 0, \theta < 0$$

多项式核、RBF 核、Sigmoid 核训练平面上的两类点得出的分类图像，如图 5-6 所示。

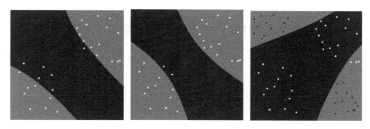

图 5-6　核函数对应图像

以下简要介绍 **Hilbert 空间**理论。

对定义于 $[\boldsymbol{a}, \boldsymbol{b}]$ 上的实函数等间隔取样，得到点列

$$(a, x_1, x_2, ... x_n, b)$$

这样可以通过向量

$$\boldsymbol{F} = (f(a), f(x_1), f(x_2), ..., f(x_n), f(b))^{\mathrm{T}}$$

近似地表征 $f(\boldsymbol{x})$。根据信号采样理论，只要采样频率大于 $f(\boldsymbol{x})$ 频谱最大值的两倍，向量 $\boldsymbol{F}$ 就是 $f(\boldsymbol{x})$，但通常向量 $\boldsymbol{F}$ 是无限维的，如图 5-7 所示。

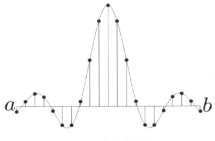

图 5-7　实函数内积

因此，根据积分定义，我们很容易将向量内积的概念拓展至两个函数的内积。

$$(f,g) = \int f(x)g(x)\mathrm{d}x$$

内积函数空间：设 H 是内积空间，其元素是某个抽象集合 B 上的实值或复值函数，则称 H 是 B 上的内积函数空间，也称 H 是 B 上的内积空间。

再生核 Hilbert 空间：设 H 是 B 上的内积函数空间，其内积为 $(.,.)$，设 $K(t,s)$ 是 B 上的二元函数，如果对每个给定的 $s \in B$，$K(t,s)$ 作为 t 的函数是 H 中的元素，且对任意的 $f \in H$，有

$$f(s) = (f(.), K(.,s))$$

则称 $K(t,s)$ 是内积空间 H 的再生核，H 称为 B 上的再生核内积函数空间。若 H 是完备的再生核内积函数空间，则称 H 为再生核 Hilbert 空间。

因此，对于上述定义的再生核 Hilbert 空间 H，映射 $\phi(x)$ 为

$$\phi(x) = K(x,.) \tag{5.12}$$

核函数 $K(t,s)$ 可以由 $\phi(x)$ 的内积形式表示为

$$K(x,y) = (\phi(x), \phi(y))$$

首先，映射 $\phi(x)$ 的含义是将欧式空间的点 x 映射为 Hilbert 空间中的一个函数；**其次**，根据线性空间理论，此空间中任何一个点（函数）均可以通过无穷个正交的基（函数）表示。

由此可以厘清脉络：在欧式空间中线性不可分的点集，在映射到无穷维的特征空间（再生核 Hilbert 空间）后线性可分；由于在我们求解最优超平面过程中只用到内积的形式，且在(5.12)式中定义的非线性变换后的特征空间维度为无穷大，我们只需要研究内积形式的核函数即可。此外，由核函数定义的 Hilbert 空间有一套相应的理论支撑，为我们找寻(5.11)式定义的高维映射提供了方向。

但是，为什么不直接使用非线性的方法分类呢？比如在平面上有两类线性不可分的点，我们为什么不直接求得一个 $y = f(x)$，使得两类点分布在这个曲线的两侧呢？笔者认为有两个原因，**其一**是我们还没有解决此类非线性问题的有效方法；**其二**是通过非线性映射转换成高维空间中的线性问题，这在现有数学框架下是可以理解的。尽管我们在研究抽象的问题上取得了很多成果，但是这些抽象来源于我们的认知实践，来源于可解释性。但是，谁能告诉我们对于某个具体的训练样本集，该选用什么样的核函数呢？

5.4.5 软间隔

机器学习的重要指标是算法的泛化能力，但是用训练学得的结果去预测未知常常会受到过拟合的困扰，因为我们需要的是经验背后的共性，或者说是经验的深度抽象。部分机器学习算法的基本思想是以训练集上的误差（也被称为经验风险）最小化代替泛化误差的最小化，但在训练样本有限时算法的泛化能力较差。为此，统计学习理论权衡经验风险和置信区间以取得实际风险最小化，最优超平面就是按照这个逻辑设计的。而软间隔算法是在综合考虑经验风险和结构风险的基础上，设计出的机器学习算法。

考虑在最优超平面的基础上增加少量的干扰点，这些干扰点分布在最优隔离带中间，可能出现两种情况，一种是这些点没有越过最优超平面；另一种是有些点越过超平面使得训练集线性不可分。对于前者可以重新训练，减小最优隔离带的长度；对于后者可以使用核方法进行高维特征映射，但是这都会因为少量的扰动带来过拟合。因此，我们考虑加入误分类样本点数量以降低经验误差。图 5-8 所示为这两种情况的示意图。

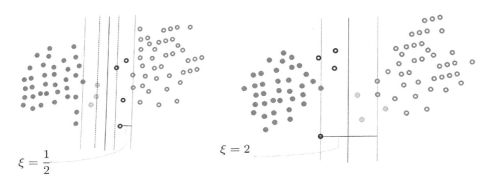

图 5-8 软间隔

对每个训练样本点，定义如下损失量：

$$\ell_{0/1}(y_i(a^T x_i - b) - 1)$$

其中

$$\ell_{0/1}(z) = \begin{cases} 1, & \text{若} \quad z < 0 \\ 0, & \text{否则} \end{cases}$$

由于上述函数的不连续性质会给后续的分析带来困难，我们寻找单调的凸函数代替，并定义松弛变量 ξ 表征训练样本点距离最优隔离带本方边缘的距离，则最优超平面的问题等价为

$$\mathop{\mathrm{argmin}}_{(a,b,\xi)} \{\frac{1}{2}\|a\|^2 + CF(\sum_{i=1}^{m}\xi_i^\sigma)\}$$

当 $\quad \sigma > 0$
$\quad\quad F : F(0) = 0$
$\quad\quad C : 常量$
s.t. $\quad y_i(a^T x - b) \geqslant 1 - \xi_i$
$\quad\quad \xi_i \geqslant 0$

取

$$F(u) = u^k, k > 1, \sigma = 1$$

则拉格朗日对偶函数为

$$L(\boldsymbol{a}, b, \boldsymbol{\xi}, \boldsymbol{\Lambda}, \boldsymbol{R}) = \frac{1}{2}\boldsymbol{a}^\mathrm{T}\boldsymbol{a} + C(\sum_{i=1}^{m}\xi_i)^k + \sum_{i=1}^{m}\lambda_i(1 - \xi_i - y_i(\boldsymbol{a}^\mathrm{T}\boldsymbol{x}_i - b)) + \sum_{i=1}^{m}\gamma_i(-\xi_i)$$

其中

$$\boldsymbol{R} = (\gamma_1, \gamma_2, ..., \gamma_m)^\mathrm{T}$$

令原求解域的梯度为 0,得

$$\frac{\partial}{\partial \boldsymbol{a}}L(\boldsymbol{a}, b, \boldsymbol{\xi}, \boldsymbol{\Lambda}, \boldsymbol{R}) = \boldsymbol{a} + \sum_{i=1}^{m}\lambda_i(0 - y_i(\boldsymbol{x}_i - 0)) \qquad = 0$$

$$\frac{\partial}{\partial b}L(\boldsymbol{a}, b, \boldsymbol{\xi}, \boldsymbol{\Lambda}, \boldsymbol{R}) = 0 + \sum_{i=1}^{m}\lambda_i(0 - y_i(0 - 1)) \qquad = 0$$

$$\frac{\partial}{\partial \boldsymbol{\xi}_i}L(\boldsymbol{a}, b, \boldsymbol{\xi}, \boldsymbol{\Lambda}, \boldsymbol{R}) = kC(\sum_{i=1}^{m}\xi_i)^{k-1} + \sum_{i=1}^{m}\lambda_i(-1) + \sum_{i=1}^{m}\gamma_i(-) = 0$$

令

$$kC(\sum_{i=1}^{m}\xi_i)^{k-1} = \delta$$

则

$$L(\boldsymbol{a}, b, \boldsymbol{\xi}, \boldsymbol{\Lambda}, \boldsymbol{R}) = \sum_{i=1}^{m}\lambda_i - \frac{1}{2}\sum_{i=1}^{m}\sum_{j=1}^{m}\lambda_i\lambda_j y_i y_j \boldsymbol{x}_i^\mathrm{T}\boldsymbol{x}_j - \frac{\delta^{k/k-1}}{(kC)^{1/k-1}}(1 - \frac{1}{k})$$

$$= \boldsymbol{1}\boldsymbol{\Lambda} - \frac{1}{2}\boldsymbol{\Lambda}^\mathrm{T}\boldsymbol{P}\boldsymbol{\Lambda} - \frac{\delta^{k/k-1}}{(kC)^{1/k-1}}(1 - \frac{1}{k})$$

取 $k = 1$,则问题的最终形式为

$$\begin{aligned}
\min \quad & W(\Lambda) = -\Lambda^T \mathbf{1} + \frac{1}{2}\Lambda^T P \Lambda \\
\text{s.t.} \quad & \Lambda^T y = 0 \\
& \Lambda - C\mathbf{1} \preceq 0 \\
& -\Lambda \preceq 0
\end{aligned}$$

形式总是那么统一，那么优美。这种形式的软间隔 SVM 被称为 C-SVC（C-Support Vector Classification）。

现在，我们综合软间隔和核方法于最优超平面，得到对偶问题。

$$\begin{aligned}
\min \quad & W(\Lambda) = -\Lambda^T \mathbf{1} + \frac{1}{2}\Lambda^T P \Lambda \\
\text{s.t.} \quad & \Lambda^T y = 0 \\
& \Lambda - C\mathbf{1} \preceq 0 \\
& -\Lambda \preceq 0
\end{aligned} \tag{5.13}$$

其中

$$\begin{aligned}
P_{ij} &= y_i y_j K(\boldsymbol{x}_i, \boldsymbol{x}_j) \\
\boldsymbol{y} &= (y_1, y_2, ..., y_m)^T \\
\Lambda &= (\lambda_1, \lambda_2, ..., \lambda_m)^T
\end{aligned}$$

5.4.6 优化解法

在思考题中，我们对优化解法进行探索，我们的出发点是想通过迭代训练样本集的以避免凸优化数值求解，因为凸优化的维度为训练样本数，这导致优化的维度过高，但是这种想法似乎不能成立。基于固定大部分变量求解的思想，Osuna 等创造了分块理论，简化了对偶问题的求解，其核心思想是利用凸函数及 KKT 条件分割 Λ，这样每次迭代只优化部分变量（如两个变量 λ_i 和 λ_j）。

所谓分治算法，就是当待求解的问题异常复杂、难于直接求解时，可以将此问题分解成多个子问题，所有子问题的解组合成整个问题的解。

而分块理论，是这个思想的一个实践。

假设一个球体砍掉上半部分，下半部分稳定地放置在三维空间的圆点，那么如何求解这个全局最小点呢（当然，假定你并不以解析的方式求解这个三维空间曲面的全局极值）？我们先固定 x，比如令 $x = x_0$，这样得到一个二维曲线，在这个二维曲线上求解极值，得到

$$x = x_0, y = 0, z = z_0$$

然后，固定 y，令 $y = 0$，求解极值得到

$$x = 0, y = 0, z = 0$$

这时，我们可通过梯度等方法判断当前是极值点，且是全局极值点，可借助图 5-9 所示的内容形象地理解。

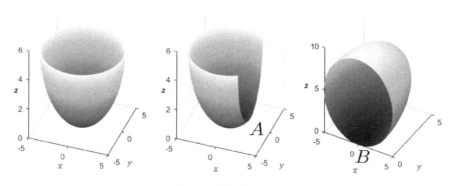

图 5-9 优化解法

接下来，我们进行数学推导。

对(5.13)式进一步实施拉格朗日对偶：

令

$$\boldsymbol{\Upsilon}^{\mathrm{T}} = (\upsilon_1, \upsilon_2, ..., \upsilon_m)$$
$$\boldsymbol{\Pi}^{\mathrm{T}} = (\pi_1, \pi_2, ..., \pi_m)$$

则
$$L(\Lambda, \mu, \Upsilon, \Pi) = W(\Lambda) + \mu\Lambda^T Y + \Upsilon^T(\Lambda - C\mathbf{1}) + \Pi^T(-\Lambda)$$
$$\nabla L(\Lambda, \mu, \Upsilon, \Pi) = -\mathbf{1} + P\Lambda + \mu Y + \Upsilon^T - \Pi^T = 0$$

求得限制条件为

$$\begin{aligned} -\mathbf{1} + P\Lambda + \mu Y + \Upsilon^T - \Pi^T &= 0 \\ \Upsilon^T(\Lambda - C\mathbf{1}) &= 0 \\ \Pi^T \Lambda &= 0 \\ \Lambda^T Y &= 0 \\ \Lambda - C\mathbf{1} &\preceq 0 \\ -\Lambda &\preceq 0 \end{aligned} \quad (5.14)$$

其中

$$f(\boldsymbol{x}) = \sum_{i=1}^{m} \lambda_i y_i K(\boldsymbol{x}_i, \boldsymbol{x}) - b$$

由(5.14)式可得到关于训练样本的最优条件（Optimality Condition）：

$$\begin{aligned} &1. \quad 0 < \lambda_i < C, \quad y_i f(\boldsymbol{x}_i) = 1 \\ &2. \quad \lambda_i = C, \quad\quad\; y_i f(\boldsymbol{x}_i) \leqslant 1 \\ &3. \quad \lambda_i = 0, \quad\quad\; y_i f(\boldsymbol{x}_i) \geqslant 1 \end{aligned} \quad (5.15)$$

设计分块（Chunk）算法，首先拆分Λ为两部分，即$\{\Lambda_B, \Lambda_N\}$，固定Λ_N为常量，只求解Λ_B代表的子优化问题。

（1）初始化Λ_B，即选取Λ的某部分作为待优化变量。

（2）根据凸优化算法，求解Λ_B代表的子优化问题。对(5.13)式进行拆分：

$$\begin{aligned}
\min \quad W(\Lambda) &= -\Lambda^T 1 + \frac{1}{2}\Lambda^T P \Lambda \\
&= -\Lambda_B^T 1 + \frac{1}{2}[\Lambda_B^T P_{BB}\Lambda_B + \Lambda_N^T P_{NN}\Lambda_N + \Lambda_B^T P_{BN}\Lambda_N + \Lambda_N^T P_{NB}\Lambda_B] - \Lambda_N^T 1 \\
&= -\Lambda_B^T 1 + \frac{1}{2}[\Lambda_B^T P_{BB}\Lambda_B\Lambda_N + \Lambda_B^T P_{BN}\Lambda_N + \Lambda_N^T P_{NB}\Lambda_B] + \{\frac{1}{2}\Lambda_N^T P_{NN} - \Lambda_N^T 1\}
\end{aligned}$$

上式后面的部分对子优化问题来说是常量，因此上式可等价如下：

$$\min \quad W'(\Lambda) = -\Lambda_B^T 1 + \frac{1}{2}[\Lambda_B^T P_{BB}\Lambda_B\Lambda_N + \Lambda_B^T P_{BN}\Lambda_N + \Lambda_N^T P_{NB}\Lambda_B]$$

$$\begin{aligned}
\text{s.t.} \quad & \Lambda_B^T y_B + \Lambda_N^T y_N = 0 \\
& \Lambda_B - C1 \preceq 0 \\
& -\Lambda_B \preceq 0
\end{aligned}$$

（3）验证Λ_N中不满足(5.15)式的条件样本点对应的λ_j，然后随意选择Λ_B中的点，如λ_i，交换两个变量，重复步骤2。

当Λ_B中只有两个变量时，即为 SMO 算法，这个算法的突出特点是可通过解析的方法求解，规避了凸优化的数值方法。

5.4.7　SVM 的 FlinkML 实现

1. 核心思想

SVM 的 FlinkML 实现的核心思想有两个。

（1）SDCA（坐标上升，Stochastic Dual Coordinate Ascent），即假定Λ_B中只有一个变量，运用优化算法求解以确定选取的变量在本次优化后的最优值，这样一次优化可以选择Λ中的一个变量，而一个迭代过程可以随机选择多个变量（位置）。

FlinkML 使用 hinge 损失函数代替$\ell_{0/1}$，

$$\ell(z) = \max(0, 1-z)$$

而且由下式可知a和Λ存在线性关系，

$$a = \sum_{i=1}^{m} \lambda_i y_i x_i$$

因此，可由对偶问题的 $\nabla \Lambda$ 求出原问题的 ∇a。CoCoA 算法利用了这个关系。

此外，考虑到 Λ 并不是独立的，即 $\Lambda^T y = 0$，如果仅仅固定其中一个变量，则此变量可以被其他变量导出。而且两个变量的优化算法可以求得最优超平面的闭式解，因此 SDCA 是 SMO 算法的简化实现。但是，在 CoCoA 算法中，数据被分成多个块（类似于 SVM 优化解法），对于每一块数据（比如第 k 块），则 $\Lambda_k^T y_k \neq 0$，因此 SDCA 选取一个变量进行优化的方法是合理的。

（2）CoCoA（Communication-efficient distributed dual Coordinate Ascent），在分布式计算环境下求解 SVM，基本原理仍然是 SDCA 算法。

数据分区增强了本地计算的优势，减少了数据传输通信开销。CoCoA 假定有 K 个机器并行地执行本地 SDCA 优化，在每一次迭代（t）的优化过程中，每个机器 k 对 a 调整 ∇a_k，且设定分区参数 β，则在每一次迭代后全局 ∇a 计算公式为

$$a^{(t)} \leftarrow a^{(t-1)} + \frac{\beta}{K} \sum_{k=1}^{K} \nabla a_k$$

上式解释如下：

（1）K 在每一次迭代中只产生一次广播通信（JobManager）以便下发新的 a。

（2）由于 a 和 Λ 存在线性关系，对于每个机器的每一次迭代需要全局的权重（即 a）参与计算。

（3）在待优化的对偶函数 $W(\Lambda)$ 中可以加上正则化，即二次项再乘以一个系数，在前面理论推导过程中没有加上正则化（即默认参数值为 1.0）。

相应地，SVM 有两个核心的实现方法，即 localDualMethod 和 fit 方法。为了理解 FlinkML 程序，需要注意符号的含义：

（1）在算法推导过程中，我们用 a 表示权重，而在程序中我们用 w 表示权重。

（2）在算法推导过程中，对偶中的拉格朗日乘子用λ表示，在程序中的以 Alphas 表示。

（3）定义$\frac{\beta}{K}$为 scaling。

2. localDualMethod

接口参数的含义如下：

（1）本地方法需要全局权重（矩阵用 DataSet[BreezeDenseVector[Double]]表示）。

（2）blockedInputNumberElements：分块后的全量训练数据。

（3）localIterations：一个迭代过程需要迭代的次数。

（4）regularization：正则化系数。

（5）scaling：$\frac{\beta}{K}$。

（6）seed：随机数种子，用于随机选择 SDCA 一次优化的变量（位置）。

（7）返回为权重矩阵的 delta 值，即∇a_k。

```
private def localDualMethod(
  w: DataSet[BreezeDenseVector[Double]],
  blockedInputNumberElements: DataSet[(Block[LabeledVector], Int)],
  localIterations: Int,
  regularization: Double,
  scaling: Double,
  seed: Long)
: DataSet[BreezeDenseVector[Double]] = {
```

方法是使用 RichMapFunction 实现自定义的 map 函数，并在 DataSet 上调用 map 操作，可以看出这里使用的是 DataSet API。

```
blockedInputNumberElements.map(localSDCA).withBroadcastSet(w,
WEIGHT_VECTOR_BROADCAST_NAME)
```

深入 map 之前，先弄清楚数据分块和分区两个概念的区别与联系。FlinkML 提供将数据分成多个块（Block）的方法。

```
case class Block[T](index: Int, values: Vector[T])

def block[T: TypeInformation: ClassTag](
  input: DataSet[T],
  numBlocks: Int,
  partitionerOption: Option[Partitioner[Int]] = None)
: DataSet[Block[T]] = {
```

在理论推导中，我们知道原求解问题的维度（a）也许并不高，但是对偶问题的维度通常很高（Λ），因为后者的维度由训练数据集的长度决定，一个训练样本增加一个维度。维度太大往往导致优化很难开展，降维是 CoCoA 的核心思想之一，所以分块（如优化解法中的分块）是从机器学习问题优化求解的角度引入的，而分区是用来解决分布式计算性能问题的。一个分区可能包括多个分块，且 CoCoA 算法中的 SDCA 是针对每一块的，而不是分区。

每一个 map 实例处理一个数据分区，用 idMapping 保存本实例要处理的所有块号（block.index），并记录本次处理块的位置（localIndex）。其中 map 的元素形式为(Block[LabeledVector], Int)，第一个属性域为一块训练数据集，第二个属性域为全部训练数据集的大小，即全局对偶问题的维度（不是每一个数据块的长度）：

```
override def map(blockNumberElements: (Block[LabeledVector], Int))
: BreezeDenseVector[Double] = {
 val (block, numberElements) = blockNumberElements
 val localIndex = idMapping.get(block.index) match {
  case Some(idx) => idx
  case None =>
   idMapping += (block.index -> counter)
   counter += 1
   alphasArray += BreezeDenseVector.zeros[Double](block.values.length)
   counter - 1
 }
```

接下来，在一个块中随机选择一个位置，运行一次优化算法 SDCA。一次优化迭代过程需要多次运行优化 SDCA，需要累积每一次 SDCA 过程中权重的更新情况，且需要将更新后的权重回传下一次 SDCA 过程。本块对应的拉格朗日乘子更新后的值也要全局保存，下一次对本块做 SDCA 迭代需要在这个值的基础上继续。其中，maximize 是一次 SDCA 优化过程，其实现原理较为简单，分析过程留给读者，代码如下：

```
val tempAlphas = alphasArray(localIndex).copy
val numLocalDatapoints = tempAlphas.length
val deltaAlphas = BreezeDenseVector.zeros[Double](numLocalDatapoints)
val w = originalW.copy
val deltaW = BreezeDenseVector.zeros[Double](originalW.length)
for(i <- 1 to localIterations) {
  // pick random data point for SDCA
  val idx = r.nextInt(numLocalDatapoints)
  val LabeledVector(label, vector) = block.values(idx)
  val alpha = tempAlphas(idx)
  // maximize the dual problem and retrieve alpha and weight vector updates
  val (deltaAlpha, deltaWUpdate) = maximize(
    vector.asBreeze,
    label,
    regularization,
    alpha,
    w,
    numberElements)
  // update alpha values
  tempAlphas(idx) += deltaAlpha
  deltaAlphas(idx) += deltaAlpha
  // deltaWUpdate is already scaled with 1/lambda/n
  w += deltaWUpdate
  deltaW += deltaWUpdate
}
// update local alpha values
alphasArray(localIndex) += deltaAlphas * scaling
deltaW
```

3. fit

fit 根据并行度或超参设定分块数量，并初始化权重为 0：

```
val blocks = resultingParameters.get(Blocks) match {
  case Some(value) => value
  case None => input.getParallelism
```

```
}

val initialWeights = createInitialWeights(dimension)

val blockedInputNumberElements = FlinkMLTools.block(
  input,
  blocks,
  Some(ModuloKeyPartitioner)).
  cross(numberVectors).
  map { x => x }
```

然后，创建固定数量的迭代器（CoCoA 迭代），更新全局权重值：

```
val resultingWeights = initialWeights.iterate(iterations) {
  weights => {
    // compute the local SDCA to obtain the weight vector updates
    val deltaWs = localDualMethod(
      weights,
      blockedInputNumberElements,
      localIterations,
      regularization,
      scaling,
      seed
    )

    // scale the weight vectors
    val weightedDeltaWs = deltaWs map {
      deltaW => {
        deltaW :*= scaling
      }
    }

    // calculate the new weight vector by adding the weight vector updates to the weight
    // vector value
    weights.union(weightedDeltaWs).reduce { _ + _ }
  }
}
```

4．超参设置

在实现类中定义了一系列超参，包括训练数据分块数、外层迭代次数、内层迭代次数、正则化系数、scaling、判断收敛的条件值等，代码如下：

```
class SVM extends Predictor[SVM] {

  import SVM._

  /** Stores the learned weight vector after the fit operation */
  var weightsOption: Option[DataSet[DenseVector]] = None

  /** Sets the number of data blocks/partitions
    *
    * @param blocks the number of blocks into which the input data will be split.
    * @return itself
    */
  def setBlocks(blocks: Int): SVM = {
    parameters.add(Blocks, blocks)
    this
  }

  /** Sets the number of outer iterations
    *
    * @param iterations the maximum number of iterations of the outer loop method
    * @return itself
    */
  def setIterations(iterations: Int): SVM = {
    parameters.add(Iterations, iterations)
    this
  }

  /** Sets the number of local SDCA iterations
    *
    * @param localIterations the maximum number of SDCA iterations
    * @return itself
    */
  def setLocalIterations(localIterations: Int): SVM = {
    parameters.add(LocalIterations, localIterations)
    this
  }

  /** Sets the regularization constant
    *
    * @param regularization the regularization constant of the SVM algorithm
```

```
 * @return itself
 */
def setRegularization(regularization: Double): SVM = {
  parameters.add(Regularization, regularization)
  this
}

/** Sets the stepsize for the weight vector updates
  *
  * @param stepsize the initial step size for the updates of the weight vector
  * @return itself
  */
def setStepsize(stepsize: Double): SVM = {
  parameters.add(Stepsize, stepsize)
  this
}

/** Sets the seed value for the random number generator
  *
  * @param seed the seed to initialize the random number generator
  * @return itself
  */
def setSeed(seed: Long): SVM = {
  parameters.add(Seed, seed)
  this
}

/** Sets the threshold above which elements are classified as positive.
  *
  * The [[predict ]] and [[evaluate]] functions will return +1.0 for items with a decision
  * function value above this threshold, and -1.0 for items below it.
  * @param threshold the limiting value for the decision function above which examples are
  *                  labeled as positive
  * @return itself
  */
def setThreshold(threshold: Double): SVM = {
  parameters.add(ThresholdValue, threshold)
  this
}

/** Sets whether the predictions should return the raw decision function value or the
  * thresholded binary value.
```

```
 *
 * When setting this to true, predict and evaluate return the raw decision value, which is
 * the distance from the separating hyperplane.
 * When setting this to false, they return thresholded (+1.0, -1.0) values.
 *
 * @param outputDecisionFunction When set to true, [[predict]] and [[evaluate]] return the raw
 *                       decision function values. When set to false, they return the
 *                       thresholded binary values (+1.0, -1.0).
 * @return itself
 */
def setOutputDecisionFunction(outputDecisionFunction: Boolean): SVM = {
  parameters.add(OutputDecisionFunction, outputDecisionFunction)
  this
}
}
```

在 SVM 的伴生对象中设置这些参数的默认值，代码如下：

```
object SVM{

val WEIGHT_VECTOR_BROADCAST_NAME = "weightVector"

case object Blocks extends Parameter[Int] {
  val defaultValue: Option[Int] = None
}

case object Iterations extends Parameter[Int] {
  val defaultValue = Some(10)
}

case object LocalIterations extends Parameter[Int] {
  val defaultValue = Some(10)
}

case object Regularization extends Parameter[Double] {
  val defaultValue = Some(1.0)
}

case object Stepsize extends Parameter[Double] {
  val defaultValue = Some(1.0)
}
```

```
case object Seed extends Parameter[Long] {
  val defaultValue = Some(Random.nextLong())
}

case object ThresholdValue extends Parameter[Double] {
  val defaultValue = Some(0.0)
}

case object OutputDecisionFunction extends Parameter[Boolean] {
  val defaultValue = Some(false)
}
```

5.4.8　SVM 的应用

SVM 应用需要主动设置超参。

- Blocks：训练数据分为多少块。
- Iterations：外层迭代次数。
- LocalIterations：内层迭代次数，即 localDualMethod 的迭代次数。
- Regularization：正则化系数，默认值为 1.0。
- Stepsize，即 β。
- ThresholdValue：判断收敛的条件值，默认为 0.0，所以必须事先设定。
- OutputDecisionFunction：即标签值的范围，如二分类任务以 1.0 代表正类，-1.0 代表负类。

样例代码如下：

```
// Create the SVM learner
val svm = SVM()
  .setBlocks(10)

// Learn the SVM model
svm.fit(trainingDS)

// Calculate the predictions for the testing data set
val predictionDS: DataSet[(Vector, Double)] = svm.predict(testingDS)
```

5.5 推荐算法

信息检索的典型解决方式是分类和搜索，如谷歌搜索引擎。这两种解决方式需要用户提供明确的需求，如检索的目录结构、搜索关键词等，而推荐系统通过分析用户的行为特征主动推送信息，推荐算法是其关键。

5.5.1 推荐系统的分类

推荐的本质是建立用户和物品之间的联系，主要解决以下两个问题。

（1）单一物品的喜好估计。

（2）多个物品的预测排名。

第一个问题是由用户发起的，主要作用是精准预测用户喜好某个物品的程度；第二个问题则由推荐系统解决，提供该用户喜好物品列表的 top-N 排名。根据推荐策略的不同，可以将推荐系统划分为以下三类。

1. 协同过滤（CF，Collaborative Filtering）推荐系统

协同过滤算法是推荐系统中应用最多的算法，其核心思想是假定用户的喜好特征是静态的，现在的喜好延续了过去的喜好，并在此假定的基础上根据相关人群的喜好，而不是根据单个物品的特征推荐。这不难理解，通常我们的购物决定是通过周围朋友的推荐而做出的，因为在潜意识里我们相信集体智慧是最强大的。

如图 5-10 所示，假定系统有 m 个用户，n 个物品，则推荐系统需要构建一个 $m \times n$ 用户物品矩阵 R（User-Item Matrix）。矩阵的项 $r_{i,j}$ 代表用户 u_i 喜好物品 p_j 的程度（或者称为评价），因此行向量 R_i 代表全体物品对于用户 u_i 的喜好权重向量。由于矩阵的行列值通常很大，而且矩阵是稀疏的，协同过滤等价于预测用户 a 是否喜好物品 q，或者给出一个 top-N 物品列表。

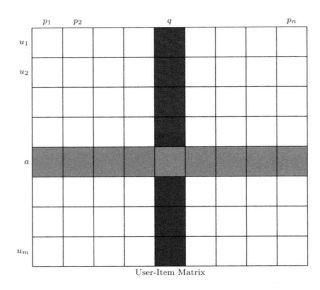

图 5-10 协同过滤

协同过滤主要有两种算法实现形式。

（1）按照行或者列计算向量间的相似度。计算行向量间相似度的形式被称为基于用户的协同过滤，计算列向量间相似度的形式被称为基于物品的协同过滤，下面以基于用户的协同过滤为例进行阐释。

根据行向量（将用户对所有物品的偏好作为向量）计算用户间的相似度，找到 K 个近邻（kNN，k-Nearest Neighbor）后，根据近邻的相似度权重及它们对物品的偏好，预测当前用户偏好未涉及的物品（即行向量稀疏的部分）。

（2）训练 CF 任务模型，如贝叶斯、聚类、决策树、矩阵奇异值分解（SVD，Singularly Valuable Decomposition）等。

2. 基于内容（CB，Content-Based）的推荐系统

基于内容的推荐系统根据物品的特征属性（Descriptive Attributes）和用户的喜好特征（profile）计算出推荐物品，核心思想是推荐与用户过去喜好相似的物品。在电商平台购物时，平台常常推荐与购买物品相关的一系列产品，这就是基

于内容的推荐系统的应用。基于内容的推荐系统通常由三个部分组成：物品特征抽取、用户喜好特征学习及推荐计算。

这类推荐系统利用用户和物品的个体特征进行匹配分析，因此用户之间是相互独立的，但是物品特征抽取通常比较困难，也无法为新用户生成推荐。

3. 混合（Hybrid）推荐系统

可以组合协同过滤和基于内容的推荐系统为混合推荐系统，以同时发挥两者的优点。

此外，基于深度学习的方法也成功应用于推荐系统中，如基于受限玻尔兹曼机（RBM，Restricted Boltzmann Machine）、深度信念网络（DBN，Deep Belief Network）、自编码器（AutoEncoder）、循环神经网络（RNN，Recurrent Neural Network）、卷积神经网络（CNN，Convolutional Neural Network）等方法，本节不再详述。

借助以上深度学习的方法，推荐算法能够在提升推荐的精准度、应对冷启动、分布式应用方面给出不同于传统方法的解决方案。近年来，随着深度学习的大潮来临，推荐算法成为研究的热点。

5.5.2　ALS-WR 算法

根据奇异值分解理论，任意实矩阵 $\boldsymbol{R} \in \mathbb{R}^{m \times n}$ 可以表示为

$$\boldsymbol{R} = \boldsymbol{U}\boldsymbol{\Sigma}\boldsymbol{M}^{\mathrm{T}}$$
$$\boldsymbol{U} \in \mathbb{R}^{m \times m}$$
$$\boldsymbol{M} \in \mathbb{R}^{n \times n}$$

其中 $\boldsymbol{U}$ 和 $\boldsymbol{M}$ 是酉矩阵，$\boldsymbol{\Sigma}$ 是半正定对角矩阵，其对角线上的元素为 $\boldsymbol{R}$ 的奇异值，$\boldsymbol{U}$ 和 $\boldsymbol{M}$ 的列向量是 $\boldsymbol{R}$ 的特征向量。基于特征分解的思想，正则化交替最小二乘法（ALS-WR，Alternating Least Squares with Weighted Regularization）常用于求解协同过滤推荐问题。

我们只可能收集到用户对部分物品的评价（购买）信息，所以协同过滤推荐系统中用户物品矩阵 $\boldsymbol{R}$ 的大部分值是未知的，因此问题转换为用矩阵中已知的值估计未知值，低秩矩阵近似是这个问题的解决方案。

用 $\tilde{\boldsymbol{R}} = \boldsymbol{U}^{\mathrm{T}}\boldsymbol{M}$ 近似 $\boldsymbol{R}$，其中

$$\boldsymbol{U} = [\boldsymbol{u}_i], \qquad \boldsymbol{u}_i \subseteq \mathbb{R}^{n_f}$$
$$\boldsymbol{M} = [\boldsymbol{m}_j], \qquad \boldsymbol{m}_j \subseteq \mathbb{R}^{n_f}$$

期望的结果是 $r_{i,j} = \boldsymbol{u}_i^{\mathrm{T}}\boldsymbol{m}_j$。定义均方误差损失函数

$$\mathcal{L}^2(r, \boldsymbol{u}, \boldsymbol{m}) = (r - \boldsymbol{u}^{\mathrm{T}}\boldsymbol{m})^2$$

定义全体已知值的经验误差

$$\mathcal{L}^{\mathrm{emp}}(\boldsymbol{R}, \boldsymbol{U}, \boldsymbol{M}) = \frac{1}{k}\sum_{i,j}\mathcal{L}^2(r_{i,j}, \boldsymbol{u}_i, \boldsymbol{m}_j)$$

其中 k 为训练样本数量。由于矩阵 $\boldsymbol{R}$ 是稀疏的，为了规避过拟合，加上 $L2$ 正则化项，

$$(\boldsymbol{U}, \boldsymbol{M}) = \underset{\boldsymbol{U}, \boldsymbol{M}}{\mathrm{argmin}}(\mathcal{L}^{\mathrm{emp}}(\boldsymbol{R}, \boldsymbol{U}, \boldsymbol{M}) + \lambda(\|\boldsymbol{U}\boldsymbol{\Gamma}_U\|^2 + \|\boldsymbol{M}\boldsymbol{\Gamma}_M\|^2))$$

上式中

$$\boldsymbol{\Gamma}_U = \mathrm{diag}(n_{u_i}), \boldsymbol{\Gamma}_M = \mathrm{diag}(n_{m_j})$$

为特征值（对应于矩阵 $\boldsymbol{\Gamma}_U$ 和 $\boldsymbol{\Gamma}_M$ 的特征值）连接成的对角矩阵，在这里不再详述，在后面会将上式展开说明。

上面优化问题的解法如以下 4 步所示。

（1）初始化矩阵 $\boldsymbol{M}$，如用全部物品评价均值初始化第一行，其他值用小的随机数初始化。

（2）固定 $\boldsymbol{M}$ 求解上式，这时是关于 $\boldsymbol{U}$ 的二次型，可以用梯度下降等数值方法

求解。将上式展开为

$$f(U, M) = \sum_{i,j}(r_{i,j} - u_i^T m_j)^2 + \lambda(\sum_i n_{u_i}\|u_i\|^2 + (\sum_j n_{m_j}\|m_j\|^2)$$

（3）固定 U 求解上式。

（4）重复步骤 2 和步骤 3，已达到误差的设定标准。

类似 SVM 的 FlinkML 实现，分布式求解 ALS-WR 也采用分块的方式。在步骤 2 中，U 是分块的，但是所有并行实例在求解过程中需要全局的 M；同理，在步骤 3 中，M 是分块的，但是所有并行实例在求解过程中需要全局的 U。

5.5.3 FlinkML 实现

本节重点分析 fit 方法，有了 SVM 代码分析的基础，ALS-WR 的其他代码不难理解。

根据评价值的类型，fit 方法分为两个对应方法，即 Double 类型（fitALS）和 Int 类型（fitALSInt）的 fit 方法，Int 类型的 fit 方法在转换数据类型为 Double 后调用 fitALS 中的 fit 方法：

```
implicit val fitALSInt = new FitOperation[ALS, (Int, Int, Double)] {
  override def fit(
    instance: ALS,
    fitParameters: ParameterMap,
    input: DataSet[(Int, Int, Double)])
  : Unit = {
    val longInput = input.map { x => (x._1.toLong, x._2.toLong, x._3)}
    implicitly[FitOperation[ALS, (Long, Long, Double)]].fit(instance, fitParameters, longInput)
  }
}
```

fitALS 的数据定义为三元组，fit 方法首先将其转换成 case 类型：

```
case class Rating(user: Long, item: Long, rating: Double)
```

```
implicit val fitALS = new FitOperation[ALS, (Long, Long, Double)] {
  override def fit(
      instance: ALS,
      fitParameters: ParameterMap,
      input: DataSet[(Long, Long, Double)])
    : Unit = {
    //省略超参加载代码
    val ratings = input.map {
      entry => {
        val (userID, itemID, rating) = entry
        Rating(userID, itemID, rating)
      }
    }
```

通常，用户和物品向量的维度非常大（n_f），FlinkML 借助外部介质存储分块数据，其中 item 表征物品：

```
    val (uIn, uOut) = createBlockInformation(userBlocks, itemBlocks, ratingsByUserBlock,
      blockIDPartitioner)
    val (iIn, iOut) = createBlockInformation(itemBlocks, userBlocks, ratingsByItemBlock,
      blockIDPartitioner)
    val (userIn, userOut) = persistencePath match {
      case Some(path) => FlinkMLTools.persist(uIn, uOut, path + "userIn", path + "userOut")
      case None => (uIn, uOut)
    }
    val (itemIn, itemOut) = persistencePath match {
      case Some(path) => FlinkMLTools.persist(iIn, iOut, path + "itemIn", path + "itemOut")
      case None => (iIn, iOut)
    }
```

初始化物品 M 矩阵，这里用随机数初始全部元素值，withForward Fields 的参数定义注解如下：

```
    val initialItems = itemOut.partitionCustom(blockIDPartitioner, 0).map{
      outInfos =>
        val blockID = outInfos._1
        val infos = outInfos._2
        (blockID, infos.elementIDs.map{
```

```
      id =>
        val random = new Random(id ^ seed)
        randomFactors(factors, random)
    })
}.withForwardedFields("0")
```

在每轮迭代中调用 updateFactors 方法完成 5.5.2 节中定义的步骤 2 和步骤 3：

```
val items = initialItems.iterate(iterations) {
  items => {
    val users = updateFactors(userBlocks, items, itemOut, userIn, factors, lambda,
      blockIDPartitioner)
    updateFactors(itemBlocks, users, userOut, itemIn, factors, lambda,
blockIDPartitioner)
  }
}
```

到这里，对 *M* 的优化已经完成，可以持久化到外部存储系统：

```
val pItems = persistencePath match {
  case Some(path) => FlinkMLTools.persist(items, path + "items")
  case None => items
}
```

由于所有迭代的最后一步是更新物品特征矩阵，需要再进行一次 5.5.2 节中定义的步骤 2，使用户特征矩阵和持久化的物品特征矩阵达到同一个基准点：

```
// perform last half-step to calculate the user matrix
val users = updateFactors(userBlocks, pItems, itemOut, userIn, factors, lambda,
  blockIDPartitioner)
```

为了便于整体阅读，附如下 fitALS 代码：

```
/** Calculates the matrix factorization for the given ratings. A rating is defined as
  * a tuple of user ID, item ID and the corresponding rating.
  *
  * @return Factorization containing the user and item matrix
  */
implicit val fitALS = new FitOperation[ALS, (Long, Long, Double)] {
  override def fit(
      instance: ALS,
      fitParameters: ParameterMap,
      input: DataSet[(Long, Long, Double)])
```

```scala
    : Unit = {
    val resultParameters = instance.parameters ++ fitParameters

    val userBlocks = resultParameters.get(Blocks).getOrElse(input.count.toInt)
    val itemBlocks = userBlocks
    val persistencePath = resultParameters.get(TemporaryPath)
    val seed = resultParameters(Seed)
    val factors = resultParameters(NumFactors)
    val iterations = resultParameters(Iterations)
    val lambda = resultParameters(Lambda)

    val ratings = input.map {
      entry => {
        val (userID, itemID, rating) = entry
        Rating(userID, itemID, rating)
      }
    }

    val blockIDPartitioner = new BlockIDPartitioner()

    val ratingsByUserBlock = ratings.map{
      rating =>
        val blockID = (rating.user % userBlocks).toInt
        (blockID, rating)
    } partitionCustom(blockIDPartitioner, 0)

    val ratingsByItemBlock = ratings map {
      rating =>
        val blockID = (rating.item % itemBlocks).toInt
        (blockID, new Rating(rating.item, rating.user, rating.rating))
    } partitionCustom(blockIDPartitioner, 0)

    val (uIn, uOut) = createBlockInformation(userBlocks, itemBlocks, ratingsByUserBlock,
      blockIDPartitioner)
    val (iIn, iOut) = createBlockInformation(itemBlocks, userBlocks, ratingsByItemBlock,
      blockIDPartitioner)

    val (userIn, userOut) = persistencePath match {
      case Some(path) => FlinkMLTools.persist(uIn, uOut, path + "userIn", path + "userOut")
      case None => (uIn, uOut)
```

```scala
    }

    val (itemIn, itemOut) = persistencePath match {
      case Some(path) => FlinkMLTools.persist(iIn, iOut, path + "itemIn", path + "itemOut")
      case None => (iIn, iOut)
    }

    val initialItems = itemOut.partitionCustom(blockIDPartitioner, 0).map{
      outInfos =>
        val blockID = outInfos._1
        val infos = outInfos._2

        (blockID, infos.elementIDs.map{
          id =>
            val random = new Random(id ^ seed)
            randomFactors(factors, random)
        })
    }.withForwardedFields("0")

    // iteration to calculate the item matrix
    val items = initialItems.iterate(iterations) {
      items => {
        val users = updateFactors(userBlocks, items, itemOut, userIn, factors, lambda,
          blockIDPartitioner)
        updateFactors(itemBlocks, users, userOut, itemIn, factors, lambda, blockIDPartitioner)
      }
    }

    val pItems = persistencePath match {
      case Some(path) => FlinkMLTools.persist(items, path + "items")
      case None => items
    }

    // perform last half-step to calculate the user matrix
    val users = updateFactors(userBlocks, pItems, itemOut, userIn, factors, lambda,
      blockIDPartitioner)

    instance.factorsOption = Some((
      unblock(users, userOut, blockIDPartitioner),
      unblock(pItems, itemOut, blockIDPartitioner)))
  }
}
```

5.5.4 ALS-WR 的应用

根据前面的代码分析，ALS-WR 的超参需要手动设置。

- Blocks：训练数据分为多少块，默认为部分块。
- NumFactors：用户和物品向量的维度，即 n_f，默认为 10。
- Lambda：正则化系数，默认为 1.0。
- Iterations：最大迭代次数，默认为 10。
- TemporaryPath：临时存储目录。

训练数据是三元组，即(用户 ID, 物品 ID, 评价值)，样例代码如下：

```
val als = ALS()
.setIterations(10)
.setNumFactors(10)
.setBlocks(100)
.setTemporaryPath("hdfs://tempPath")

// Set the other parameters via a parameter map
val parameters = ParameterMap()
.add(ALS.Lambda, 0.9)

// Calculate the factorization
als.fit(inputDS, parameters)

// Calculate the ratings according to the matrix factorization
val predictedRatings = als.predict(testingDS)
```

5.6 思考题

（1）线性拟合的 FlinkML 实现中为什么使用梯度下降法而不用闭式解法？

（2）FlinkML 的 Pipeline 机制是在总结了 Scikit-learn 架构经验的基础上设计的，这种架构是优秀的，但在解决哪类机器学习任务时使会遇到困难？

（3）在 C-SVC 中前缀 C 的含义是什么？如果两类数据的分布是不均衡的（例

如正类数据的总量远大于负类数据的总量），那么问题的等价形式怎样调整比较合理？

（4）考虑用一种通过训练样本集的迭代优化解法。根据"最优解=隔离带最宽"的思想，仅有两个点的训练集，最优超平面应垂直于两点连线，如图 5-11 所示。

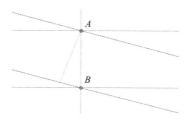

图 5-11　有两个点的训练集

三个点的训练集，应该在三角形最短边长方向上构建隔离带，如图 5-12 所示。

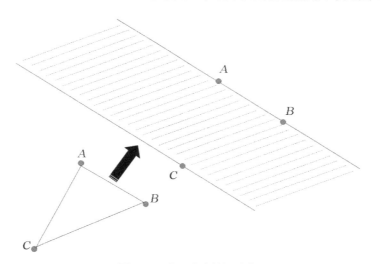

图 5-12　有三个点的训练集

根据这些推理，我们可以设计如下算法。

（1）选取三个样本构建隔离带。

（2）取出一个样本，与之前三个支持向量构建隔离带。

（3）重复步骤 2，得到最终的支持向量。

其中 4 个点的隔离带构建方案，如图 5-13 所示。

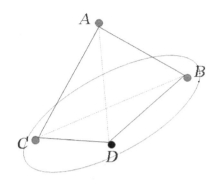

图 5-13　4 个点的隔离带构建方案

经过仔细推敲后，笔者发现上述理论不对，反例如图 5-14 所示。

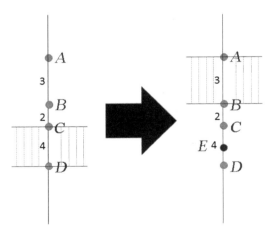

图 5-14　反例

图 5-14 说明新加入一个训练样本后的支持向量和原有的支持向量可以没有关系。为了让读者清楚上述思路，笔者绘制了整个思考过程（如图 5-15 所示）：为了避开求解 QP 问题，我们构造了一个优化算法，之后提出一个反例证明了之前的算法存在缺陷。

图 5-15　思考过程

算法和反例的构造是否合理？或者说，是否可以用一个错误去证明另一个错误？

（5）支持向量一定是最极端的样本吗？例如，在金融风控场景中，筛选客户时，支持向量一定是那些最优质的客户样本或逾期最严重的客户吗？使用 LIBSVM 训练鸢尾花数据集时，支持向量应该是什么样的训练样本呢？

第 6 章
关系型 API

关系型 API 包括 Table API 和 SQL 两类。由于这两类 API 是紧密集成在一起的，我们将两者放在本章介绍。6.1 节将介绍 Flink 引入关系型 API 的原因。为了使读者更好地理解关系型 API，6.2 节将简要介绍 SQL 解析与优化框架 Calcite；6.3 节将介绍关系型 API 的主要内容；在有界数据集上构架的 Table API 与 SQL 和关系型数据库查询非常相似，但是在 DataStream 上的构架则不易理解，为此 6.4 节将介绍动态表的相关知识。

6.1 为什么需要关系型 API

我们可以借助 DataStream API 和 DataSet API 解决复杂的数据处理问题，巧妙地构架机器学习引擎，但这需要开发人员掌握 Java 或 Scala 语言，并精通两套 API。此外，随着流处理应用的广泛推广及 Flink 在该领域的快速增长，Flink 社区决定提供同时支持批处理和流处理任务的统一的、更简单的 API，以降低开发门槛，满足更多用户的使用需求，结构化查询语言 SQL 是最好的选择。

关系型 API 是 Table API 和 SQL 的统称，其中 Table API 提供类似 LINQ（集成查询语言）语法的 API，而 SQL 则支持标准 SQL 查询语言，这类 API 有以下三个好处。

（1）它是声明式的，用户不需要提供计算的实现细节，系统决定如何计算。

（2）关系型 API 需要相应的执行引擎，起到 SQL 翻译的作用，因此我们可以在引擎层全局优化查询，而不是在编程语言层优化程序。DataStream API 和 DataSet API 应用程序则很难做到这一点。

（3）SQL 的普及程度远高于应用编程语言，尤其是在数据分析领域。

Flink 并没有重复造轮子，而是根据 SQL 解析与优化框架 Calcite 构建关系型 API。而且，尽管 SQL 是 Table API 的更高层的抽象，且尽管同一类操作的语义是相同的，但是 Table API 和 SQL 并没有被拆分成两层，应用程序可以同时使用 Table API、SQL、DataStream API 和 DataSet API。

6.2 Calcite

Apache Calcite 是为不同计算平台和数据源提供统一动态数据管理服务的高层框架，Calcite 在各种数据源上构建标准的 SQL 语言，并提供多种查询优化方案。此外，Calcite 引擎也适用于流处理场景。

Calcite 的前身为 optiq，optiq 起初在 Hive 项目中，为 Hive 提供基于模型的优化（CBO，Cost Based Optimization）。2014 年 5 月，optiq 从 Hive 中独立出来，成为 Apache 社区的孵化项目，2014 年 9 月正式更名为 Calcite。

Calcite 的目标是为不同计算平台和数据源提供统一的查询引擎（一种方案适应所有需求场景，one size fits all），并以类 SQL 查询语言访问不同数据源。

Apache Calcite 执行 SQL 查询的主要步骤如下：

（1）将 SQL 解析成未经校验的抽象语法树（AST，Abstract Syntax Tree），抽象语法树是和语言无关的形式，这和交叉编译的第一步类似。

（2）Validate：验证 AST，主要验证 SQL 语句是否合法，验证后的结果是 RelNode 树。

（3）Optimize：优化 RelNode 树并生成物理执行计划。

（4）Execute：将物理执行计划转换成特定平台的执行代码，如 Flink 的 DataStream 应用程序。

6.3 关系型 API 概述

关系型 API 是以库的形式提供的，在 Maven 工程中需要加入以下依赖：

```xml
<properties>
    <flink.version>1.6.1</flink.version>
    <scala.binary.version>2.11</scala.binary.version>
</properties>

<dependency>
    <groupId>org.apache.flink</groupId>
    <artifactId>flink-table_${scala.binary.version}</artifactId>
    <version>${flink.version}</version>
    <scope>compile</scope>
</dependency>
```

6.3.1 程序结构

关系型 API 统一了 DataStream API 和 DataSet API，而且 Table API 和 SQL 屏蔽了批处理与流处理的细节，因此 Table API 与 SQL 的程序结构是一致的。基于关系型 API 的数据处理应用程序包括以下 6 个步骤。

1. 获取运行时

运行时分为两类：StreamingExecutionEnvironment 和 ExecutionEnvironment，分别对应流处理和批处理程序，这与流处理应用程序或批处理应用程序是一致的：

```
// 流处理运行时
val env: StreamExecutionEnvironment =
StreamExecutionEnvironment.getExecutionEnvironment
// 批处理运行时
val env: ExecutionEnvironment = ExecutionEnvironment.getExecutionEnvironment
```

2. 获取 Table 运行时

SQL 应用程序也需要获取 Table 运行时：

```
val tableEnv = TableEnvironment.getTableEnvironment(env)
```

3. 注册表

SQL 应用程序也需要注册表，有以下三种方式：

```
tableEnv.registerTable("Orders", ...)
tableEnv.registerTableSource("Orders", ...)
tableEnv.registerExternalCatalog("Orders", ...)
```

4. 定义查询

Table 与 SQL 的查询格式是不同的：

```
// Table API 的 LINQ 查询
val orders = tableEnv.scan("Orders")
val sqlResult = orders
 .filter('cCountry === "FRANCE")
 .groupBy('cID, 'cName)
 .select('cID, 'cName, 'revenue.sum AS 'revSum)
// SQL 查询
val sqlResult = tableEnv.sqlQuery("SELECT ... FROM Orders...")
```

5. 写入 Sink

```
// 写入外部文件
sqlResult.writeToSink(...)
// 打印到控制台
print()
```

6. 启动程序

调用运行时的 execute()方法：

```
// 启动程序
env.execute("Kafka Window WordCount")
```

以下是 Table API 的整体程序结构：

```
// ***************
// STREAMING QUERY
// ***************
// val sEnv = StreamExecutionEnvironment.getExecutionEnvironment
// ***********
// BATCH QUERY
// ***********
// val bEnv = ExecutionEnvironment.getExecutionEnvironment

val env = StreamExecutionEnvironment.getExecutionEnvironment

// get a TableEnvironment
val tableEnv = TableEnvironment.getTableEnvironment(env)

// register a Table
tableEnv.registerTable("Orders", ...)
tableEnv.registerTableSource("Orders", ...)
tableEnv.registerExternalCatalog("Orders", ...)

// scan registered Orders table
val orders = tableEnv.scan("Orders")
// compute revenue for all customers from France
val sqlResult = orders
  .filter('cCountry === "FRANCE")
  .groupBy('cID, 'cName)
  .select('cID, 'cName, 'revenue.sum AS 'revSum)

// emit or convert Table
sqlResult.writeToSink(...)

// execute query
env.execute()
```

以下是 SQL 的程序结构：

```
// for batch programs use ExecutionEnvironment instead of StreamExecutionEnvironment
val env = StreamExecutionEnvironment.getExecutionEnvironment

// create a TableEnvironment
val tableEnv = TableEnvironment.getTableEnvironment(env)

// register a Table
tableEnv.registerTableSource("Orders", ...)

// Create a Table from a SQL query
val sqlResult = tableEnv.sqlQuery("SELECT ... FROM Orders...")

// emit a Table API result Table to a TableSink, same for SQL result
sqlResult.writeToSink(...)

// execute
env.execute()
```

以上将 Table API 和 SQL 程序分为两个例子进行了介绍，在实际应用中也可以将两种 API 混合编程，在后面章节中会详述。

6.3.2 Table 运行时

TableEnvironment 以目录（catalog）的形式管理 Schema 的元数据与命名空间，如表和 UDF，在 TableEnvironment 实例化时根 Schema 被创建：

```
// create a TableEnvironment
val tableEnv = TableEnvironment.getTableEnvironment(env)
```

在关系型数据库中，Schema 集合各类数据库对象，如表、视图、存储过程、索引等。而 Database 则是仓库，容纳所有 Schema。

关系型 API 统一了 DataStream API 和 DataSet API，因此 TableEnvironment 根据 DataStream 与 DataSet 分别定义获取 Table 运行时的方法：

```
abstract class TableEnvironment(val config: TableConfig) {
  ...
```

```
private[flink] def queryConfig: QueryConfig = this match {
  case _ : BatchTableEnvironment => new BatchQueryConfig
  case _ : StreamTableEnvironment => new StreamQueryConfig
  case _ => null
}
...
}
```

作为关系型 API 的核心对象，TableEnvironment 承担了如下重要任务：

（1）负责表的管理。目录管理 Schema 的元数据与命名空间。

（2）SQL 查询。Flink 规定同一查询（如 join、union）的表不能定义在不同的运行时环境中。

（3）UDF 管理。

（4）将 DataStream 或 DataSet 转换成表。

（5）管理 StreamExecutionEnvironment 对象和 ExecutionEnvironment 对象的引用。

此外，TableConfig 也有 DataStream 和 DataSet 的版本，主要包括 SQL 解析底层框架的配置（CalciteConfig）、时区、空字段检查启用开关。

```
class TableConfig {
 /**
  * Defines the timezone for date/time/timestamp conversions.
  */
 private var timeZone: TimeZone = TimeZone.getTimeZone("UTC")
 /**
  * Defines if all fields need to be checked for NULL first.
  */
 private var nullCheck: Boolean = true
 /**
  * Defines the configuration of Calcite for Table API and SQL queries.
  */
 private var calciteConfig: CalciteConfig = CalciteConfig.DEFAULT
 /**
  * Defines the default context for decimal division calculation.
```

```
 * We use Scala's default MathContext.DECIMAL128.
 */
private var decimalContext: MathContext = MathContext.DECIMAL128
/**
 * Specifies a threshold where generated code will be split into sub-function calls.
Java has a
 * maximum method length of 64 KB. This setting allows for finer granularity if necessary.
 */
private var maxGeneratedCodeLength: Int = 64000 // just an estimate

//以下是对应属性的 get 或 set 方法，略
}
```

从上述代码中可以清晰地看出统一批处理 API 和流处理 API 是以增加架构复杂度为代价的，并且关系型 API 是构建在 DataStream API 和 DataSet API 之上的，而不是重新造轮子。

6.3.3 表注册

不同于关系型数据库的是，这里的表分为输入表和输出表，应用程序在输入表上执行关系代数运算，然后将结果写入输出表，这是因为：

（1）Flink 引擎以 pull 方式获取外部数据，但不支持将数据更新到 Source，这是符合数据分析哲学的。

（2）应用程序的目标是获取关系代数运算的结果，但在大多数情况下，Schema 中并没有结果对应的数据结构描述。

因此，这里的表和关系型数据库中的非物化的视图比较相似。下面分别展开输入表和输出表的注册方法。

（1）输入表可以通过以下几种途径注册：

- 已有的表实例，如 Table API 或 SQL 查询的结果，代码如下：

```
// get a TableEnvironment
val tableEnv = TableEnvironment.getTableEnvironment(env)
```

```
// Table is the result of a simple projection query
val projTable: Table = tableEnv.scan("X").select(...)

// register the Table projTable as table "projectedX"
tableEnv.registerTable("projectedTable", projTable)
```

- TableSource：外部数据源。
- DataStream 或 DataSet。

（2）输出表由 TableSink 注册，这类似 DataStream 或 DataSet 里的 Sink。

6.3.4 TableSource 与 TableSink

类似 DataStream 和 DataSet，TableSource 连接外部数据源后还要解析数据源格式、定义 Schema、输出表的更新方式，最后注册表：

```
tableEnvironment
  // 连接外部数据源
  .connect(...)
  // 解析数据源格式
  .withFormat(...)
  // 定义 Schema
  .withSchema(...)
  // 输出表的更新方式
  .inAppendMode()
  // 注册表
  .registerTableSource("MyTable")
```

以下代码是读取 Kafka 中 Avro 格式的记录：

```
tableEnvironment
  // 连接 Kafka
  .connect(
    new Kafka()
      .version("0.10")
      .topic("test-input")
      .startFromEarliest()
      .property("zookeeper.connect", "localhost:2181")
      .property("bootstrap.servers", "localhost:9092")
  )
```

```
// 解析数据源格式
.withFormat(
  new Avro()
    .avroSchema(
      "{" +
      "   \"namespace\": \"org.myorganization\"," +
      "   \"type\": \"record\"," +
      "   \"name\": \"UserMessage\"," +
      "   \"fields\": [" +
      "       {\"name\": \"timestamp\", \"type\": \"string\"}," +
      "       {\"name\": \"user\", \"type\": \"long\"}," +
      "       {\"name\": \"message\", \"type\": [\"string\", \"null\"]}" +
      "   ]" +
      "}" +
    )
)

// 定义 Schema
.withSchema(
  new Schema()
    .field("rowtime", Types.SQL_TIMESTAMP)
      .rowtime(new Rowtime()
        .timestampsFromField("ts")
        .watermarksPeriodicBounded(60000)
      )
    .field("user", Types.LONG)
    .field("message", Types.STRING)
)

// 输出表的更新方式: append
.inAppendMode()

// 注册 Source 和 Sink
.registerTableSource("MyUserTable")
```

TableSink 的定义与 TableSource 相同,唯一的不同是注册形式不同。TableSink 的注册形式分别为 registerTableSink 和 registerTableSource,这里不再详述。

6.3.5 查询

本节不展开介绍关系型 API。以下是基于 Table API 的表查询代码结构：

```
// scan registered Orders table
val orders = tableEnv.scan("Orders")
val revenue = orders
// 过滤出国家为 FRANCE 的记录
  .filter('cCountry === "FRANCE")
// 按照 cID group
  .groupBy('cID, 'cName)
// 查询
  .select('cID, 'cName, 'revenue.sum AS 'revSum)
```

以下是基于 SQL 的表查询代码结构：

```
val revenue = tableEnv.sqlQuery("""
  SELECT cID, cName, SUM(revenue) AS revSum
  FROM Orders
  WHERE cCountry = 'FRANCE'
  GROUP BY cID, cName
  """.stripMargin)
```

6.3.6 相互转换

应用程序可以将 DataStream 或 DataSet 转换成 Table，也可以将 Table 转换成 DataStream 或 DataSet。

（1）若利用 Scala 隐式转换的能力，则需要 import scala 隐式转换扩展包。以下程序将(String, Int)转换成表的行，行的列名分别为 word 和 count。

```
import org.apache.flink.api.scala._
import org.apache.flink.table.api.scala._

val env = ExecutionEnvironment.getExecutionEnvironment
val tEnv = TableEnvironment.getTableEnvironment(env)

val input: DataSet[(String, Int)] = env.fromElements(("Hello", 2), ("Hello", 5), ("Ciao", 3))
val result = input
```

```
    // 将(String, Int)转换成表的行, 行的列名分别为 word 和 count, 注意以下程序中单引号的使用方式
    .toTable(tEnv, 'word, 'count)
    .groupBy('word)
    // 分组求 count 的平均
    .select('word, 'count.avg)
```

（2）注册表。以下是 StreamTableEnvironment 的接口形式，BatchTableEnvironment 有对应的接口形式：

```
def registerDataStream[T](name: String, dataStream: DataStream[T]): Unit = {
  checkValidTableName(name)
  registerDataStreamInternal(name, dataStream.javaStream)
}

def registerDataStream[T](name: String, dataStream: DataStream[T], fields:
Expression*): Unit = {
  checkValidTableName(name)
  registerDataStreamInternal(name, dataStream.javaStream, fields.toArray)
}
```

其中，第一种接口形式将 Stream 对象的属性映射成表的列；第二种接口形式通过域选择表达式选择要注册的列名，例子如下：

```
// register the DataStream as Table "myTable" with fields "f0", "f1"
tableEnv.registerDataStream("myTable", stream)

// register the DataStream as table "myTable2" with fields "myLong", "myString"
tableEnv.registerDataStream("myTable2", stream, 'myLong, 'myString)
```

（3）直接转换成表。仍以 StreamTableEnvironment 的接口形式为例：

```
def fromDataStream[T](dataStream: DataStream[T]): Table = {
  val name = createUniqueTableName()
  registerDataStreamInternal(name, dataStream.javaStream)
  scan(name)
}

def fromDataStream[T](dataStream: DataStream[T], fields: Expression*): Table = {
  val name = createUniqueTableName()
  registerDataStreamInternal(name, dataStream.javaStream, fields.toArray)
  scan(name)
}
```

这种接口的参数形式与表注册方式相同，两者的区别是用表注册方式可以指定表的名称，而用直接转换的方式则不需要指定。

（4）将表转换成 DataStream 或 DataSet。在 StreamTableEnvironment 时有两种转换模式。

- 追加模式，即表的行追加到 DataStream 中。
- 撤回模式，类似（第 1 章流式数据处理理论中的）撤回累加模式。撤回模式将 Stream 转换成二元组。

```
stream => (Boolean, stream)
```

其中，Boolean 属性定义是插入操作还是删除操作。

```
// Table with two fields (String name, Integer age)
val table: Table = ...

// convert the Table into an append DataStream of Row
val dsRow: DataStream[Row] = tableEnv.toAppendStream[Row](table)

// convert the Table into an append DataStream of Tuple2[String, Int]
val dsTuple: DataStream[(String, Int)] dsTuple =
 tableEnv.toAppendStream[(String, Int)](table)

// convert the Table into a retract DataStream of Row.
//   A retract stream of type X is a DataStream[(Boolean, X)].
//   The boolean field indicates the type of the change.
//   True is INSERT, false is DELETE.
val retractStream: DataStream[(Boolean, Row)] = tableEnv.toRetractStream[Row](table)
```

而将表转换成 DataSet 则相对简单：

```
val table: Table = ...

// convert the Table into a DataSet of Row
val dsRow: DataSet[Row] = tableEnv.toDataSet[Row](table)

// convert the Table into a DataSet of Tuple2[String, Int]
val dsTuple: DataSet[(String, Int)] = tableEnv.toDataSet[(String, Int)](table)
```

6.4 动态表概述

6.4.1 流式关系代数

SQL 的本质是关系代数，而关系代数是定义在有界数据集上的，因此将关系代数运算移植到 DataSet 上不会遇到理论上的障碍。而移植到 DataStream 上则显得不易理解：DataStream 应用程序处理的对象是无限延伸且在时间轴上动态变化的数据集，而流式关系代数理论并不能建立在数据库技术的基础之上，否则流处理会失去意义。

通过以下对比可以更清晰地看出流处理和关系代数间的区别。

（1）执行机制不同：根据 CAP 理论，可用性是有时限要求的，因此关系型数据库必须及时响应查询，并且所有数据对查询必须是可见的；而流处理应用一旦启动则一直运行下去，没有结束时间点。需要强调的是，在应用程序运行过程中，计算图逻辑结构并不会发生变化，即同一应用一直运行下去。

（2）计算结果的形式不同：SQL 查询结果的结构是事先知晓的，且结果大小是有限的；而流处理则一直在更新计算结果，并以流的形式发射出去，其长度是未知的。

尽管两者有着本质的区别，但是利用关系型查询语言架构流处理程序并非不可行，如某些关系型数据库提供的物化视图（Materialized View）机制。类似普通视图（View），物化视图也是由 SQL 语句定义的，区别于普通视图的是物化视图会缓存 SQL 查询结果，因此物化视图需要相应机制以确保查询结果和数据库一致，这可以被看作流式 SQL 查询。

下面将通过动态表来解释架构背后的机制。

6.4.2 动态表

之所以定义为动态表（Dynamic Table），是因为相较于关系型数据库的表，动态表的内容是不可预知且持续变化的，且对比于在关系型数据库上查询的执行过程，在动态表上的查询会产生持续查询（Continuous Query），这符合 Flink 流式数据处理编程模型：应用程序被翻译成计算图，其逻辑结构是固定的，数据处理任务是持续的。

在 DataStream 上执行 SQL 查询，处理过程如下：

（1）将 DataStream 转换成动态表。

（2）在动态表上定义持续查询。

（3）将持续查询的实时结果转换成动态表。

（4）将表示查询结果的动态表转换成 DataStream，这样应用程序就可以借助 DataStream API 进一步变换查询结果。

上述过程如图 6-1 所示。

图 6-1 动态表与持续查询

下面以用户点击行为分析为例进行说明。用户行为记录通常包括三个部分，用户名、用户点击的页面 URL 地址及点击时间，表示形式如下：

```
[
 user: VARCHAR,     // the name of the user
 cTime: TIMESTAMP,  // the time when the URL was accessed
 url: VARCHAR       // the URL that was accessed by the user
]
```

用户点击行为数据的动态表形式，如图 6-2 所示。

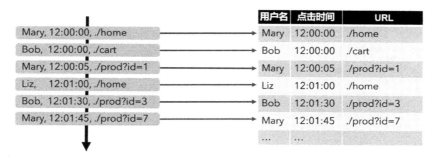

图 6-2　用户点击行为数据的动态表形式

首先，创建 ClickEvent 类，类型对应关系如表 6-1 所示。

表 6-1　用户点击行为记录类型对应关系

Table API	SQL	Java type
Types.STRING	VARCHAR	java.lang.String
Types.SQL_TIMESTAMP	TIMESTAMP(3)	java.sql.Timestamp

然后，将 DataStream[String] 转换成 DataStream[ClickEvent]，执行 SQL 查询，最后写入 Sink（文件）中，代码如下：

```scala
import java.sql.Timestamp

import org.apache.flink.api.common.typeinfo.TypeInformation
import org.apache.flink.streaming.api.scala.{StreamExecutionEnvironment, _}
import org.apache.flink.table.api.{Table, TableEnvironment, Types}
import org.apache.flink.table.sinks.CsvTableSink

// 定义ClickEvent类
case class ClickEvent(user:String, cTime:Timestamp, url:String){
  override def toString: String = user + cTime.toString + url
}

object DynamicTable {

  def main(args: Array[String]) {
    val env = StreamExecutionEnvironment.getExecutionEnvironment
    val tableEnv = TableEnvironment.getTableEnvironment(env)
```

```scala
// 将 DataStream[String] 转换成 DataStream[ClickEvent]
val input: DataStream[ClickEvent] = env.fromElements(
  "Mary,18/12/2018 12:00:00,./home",
  "Bob ,18/12/2018 12:00:00,./cart",
  "Mary,18/12/2018 12:00:05,./prod?id=1",
  "Liz ,18/12/2018 12:01:00,./home",
  "Bob ,18/12/2018 12:01:30,./prod?id=3",
  "Mary,18/12/2018 12:01:45,./prod?id=7"
).map({
  x =>
    val y: Array[String] = x.split(",")
    val ts = y(1).trim().split(" ")
    val dt = ts(0).split("/")
    val tm = ts(1).split(":")
    val ctime = new Timestamp(dt(2).toInt - 1900, dt(1).toInt -1, dt(0).toInt,
                    tm(0).toInt, tm(1).toInt, tm(2).toInt, 0)
    ClickEvent(y(0).trim(), ctime, y(2).trim())
})

// 注册表并定义查询
val Clicks: Table = tableEnv.fromDataStream(input)
tableEnv.registerTable("Clks", Clicks)
val result: Table = tableEnv.sqlQuery("SELECT * FROM Clks WHERE user='Mary'")

// 写入 CSV
val fieldNames: Array[String] = Array("user", "cTime", "url")
val fieldTypes: Array[TypeInformation[_]] = Array(Types.STRING, Types.SQL_TIMESTAMP, Types.STRING)
val csvSink = new CsvTableSink("/tmp/hai", fieldDelim = ",")
tableEnv.registerTableSink("CsvTable", fieldNames, fieldTypes, csvSink)
result.insertInto("CsvTable")

env.execute("DynamicTable")
}
}
```

程序执行的结果是在 /tmp/hai 目录下生成多个包含 DataStream 内容的文件。

6.4.3 持续查询

接下来，分析每个用户的累计点击量，SQL 语句如下：

```
SELECT user, count(url) as cnt FROM Clks GROUP BY user
```

Flink 会根据记录被观察的顺序持续执行 SQL 查询，生成动态变化的表，即表的行数会动态变化，某行的内容也会动态改变，结果如图 6-3 所示。

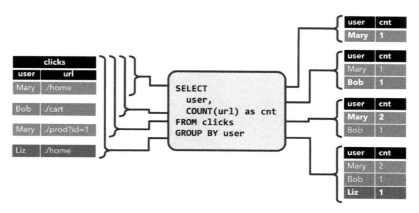

图 6-3　基于持续查询的用户点击行为分析

为了便于观察，将表转换回 DataStream 打印出来。为了按事件顺序观察，设置 map 和 print 的并行度为 1，代码如下：

```
import java.sql.Timestamp

import org.apache.flink.streaming.api.scala.{DataStream, StreamExecutionEnvironment, _}
import org.apache.flink.table.api.{Table, TableEnvironment, Types}
import org.apache.flink.types.Row

object DynamicTableWithCountUrl {
  case class ClickEvent(user:String, cTime:Timestamp, url:String){
    override def toString: String = user + cTime.toString + url
  }
  def main(args: Array[String]) {
    val env = StreamExecutionEnvironment.getExecutionEnvironment
    val tableEnv = TableEnvironment.getTableEnvironment(env)

    // 将DataStream[String]转换成DataStream[ClickEvent]
    val input: DataStream[ClickEvent] = env.fromElements(
      "Mary,18/12/2018 12:00:00,./home",
```

```
    "Bob ,18/12/2018 12:00:00,./cart",
    "Mary,18/12/2018 12:00:05,./prod?id=1",
    "Liz ,18/12/2018 12:01:00,./home",
    "Bob ,18/12/2018 12:01:30,./prod?id=3",
    "Mary,18/12/2018 12:01:45,./prod?id=7"
).map({
  x =>
    val y: Array[String] = x.split(",")
    val ts = y(1).trim().split(" ")
    val dt = ts(0).split("/")
    val tm = ts(1).split(":")
    val ctime = new Timestamp(dt(2).toInt - 1900, dt(1).toInt -1, dt(0).toInt,
                    tm(0).toInt, tm(1).toInt, tm(2).toInt, 0)
    ClickEvent(y(0).trim(), ctime, y(2).trim())
}).setParallelism(1)

// 注册表并定义查询
val Clicks: Table = tableEnv.fromDataStream(input)
tableEnv.registerTable("Clks", Clicks)
val result: Table = tableEnv.sqlQuery("SELECT user, COUNT(url) as cnt FROM Clks GROUP BY user")

/*  val appendStream: DataStream[(String, Int)] = tableEnv.toAppendStream(result)
appendStream.print()*/

// 撤回模式转换成 DataStream
val retractStream: DataStream[(Boolean, Row)] = tableEnv.toRetractStream(result)
retractStream.print().setParallelism(1)

env.execute("DynamicTable")
  }
}
```

上例中将表转换成 DataStream 采用的是撤回模式，结果如下：

```
(true,Mary,1)
(false,Mary,1)
(true,Mary,2)
```

窗口是流处理的重要机制，动态表不会丢弃这个流式数据处理的"优良基因"。以 1 小时为长度开事件时间窗口，数据保存在.cvs 文本中，在每个窗口内分析每个用户的累计点击次数，SQL 语句如下：

```
val result: Table = tableEnv.sqlQuery(
  "SELECT "+
  " user, " +
  // 开长度为 1 小时的滚动窗口
  " TUMBLE_END(cTime, INTERVAL '1' HOUR) AS endT, " +
  " COUNT(url) AS cnt " +
  "FROM Clks " +
  "GROUP BY user, " +
  // group 条件加入窗口机制，即根据窗口和用户名分组，用户名相当于 key
  "TUMBLE(cTime, INTERVAL '1' HOUR)")
```

结果如图 6-4 所示。

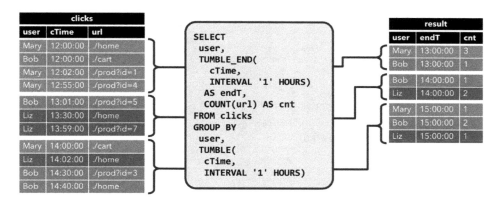

图 6-4　基于窗口查询的用户点击行为分析

完整代码如下：

```
import java.sql.Timestamp

import org.apache.flink.streaming.api.TimeCharacteristic
import org.apache.flink.streaming.api.scala.{StreamExecutionEnvironment, _}
import org.apache.flink.table.api.{Table, TableEnvironment, Types}
import org.apache.flink.table.descriptors.{Csv, FileSystem, Schema}
import org.apache.flink.types.Row

object DynamicTableCountUrlWithWindow {
  case class ClickEvent(user:String, cTime:Timestamp, url:String){
    override def toString: String = user + cTime.toString + url
```

```
}

def main(args: Array[String]) {
  val env = StreamExecutionEnvironment.getExecutionEnvironment
  env.setStreamTimeCharacteristic(TimeCharacteristic.EventTime)
  val tableEnv = TableEnvironment.getTableEnvironment(env)

  // 从文件系统中读入 DataStream
  tableEnv.connect(
    new FileSystem()
      .path("file:/// /tmp/Flink/Source")
  )
    .withFormat(
      new Csv().field("user", Types.STRING)
        .field("cTime", Types.SQL_TIMESTAMP)
        .field("url", Types.STRING)
        .fieldDelimiter(",")
        .lineDelimiter("\n")
    )
    .withSchema(
      new Schema()
        .field("user", Types.STRING)
        .field("cTime", Types.SQL_TIMESTAMP)
        .field("url", Types.STRING)
    )
    .inAppendMode()
    .registerTableSource("Clks")

  // 定义 SQL 查询
  val result: Table = tableEnv.sqlQuery(
    "SELECT " +
    " user, " +
    " TUMBLE_END(cTime, INTERVAL '1' HOUR) AS endT, " +
    " COUNT(url) AS cnt " +
    "FROM Clks " +
    "GROUP BY user, " +
    "TUMBLE(cTime, INTERVAL '1' HOUR)")

  // 追加模式转换成 DataStream
  val appendStream: DataStream[Row] = tableEnv.toAppendStream[Row](result)
  appendStream.print().setParallelism(1)
```

```
    env.execute("DynamicTable")
  }
}
```

6.5 思考题

（1）引擎会用什么方式表示动态表？

（2）应用程序可以混合 Table API、SQL、DataStream API 和 DataSet API 编程。在回答为什么需要关系型 API 时，我们指出 Flink 引擎层会提供全局查询优化，那么混入其他 API 后会对这种全局优化产生副作用吗？

（3）从动态表与持续查询的分析中可以看出，关系型 API 的本质是在处理 DataStream，从一种形式变换为另一种形式。为什么不在 DataStream API 中新增一种算子函数，而是增加一类新的 API 呢？

（4）在 JDBC 连接关系型数据库查询时，为了提高常用 SQL 语句的查询效率，先使用 PrepareStatement 方法创建 pstmt 对象，然后通过这个对象查询：

```
String sql = "select * from users where username=? and userpwd=?";
pstmt = conn.prepareStatement(sql);
pstmt.setString(1, username);
pstmt.setString(2, userpwd);
rs = pstmt.executeQuery();
```

动态表能否借用这种机制？

（5）在基于关系型 API 编写的流处理应用程序中，首先获取流处理运行时，然后获取 Table 运行时，最后通过流处理运行时的 execute() 方法触发程序被执行，但为什么不通过 Table 运行时的 execute() 方法来操作呢？

第 7 章
复杂事件处理

复杂事件处理用来检测无界数据集中的复杂模式，如受一系列事件驱动的异常检测、物联网领域中基于 RFID（Radio Frequency Identification）技术的物品追踪与监控。模式的检测过程也被称为模式匹配（Pattern Matching）。

本章先以股票异常交易检测为例，讲述模式匹配的编程过程，然后分析 Flink DataStream API 和 Table API 在解决这类问题上遇到的困难。为此，Flink 引入 NFA[b] 模式匹配编程模型，最后基于 FlinkCEP API 编程解决股票异常交易检测问题。此外，为了让读者更容易理解 NFA[b] 模型，7.2 节将介绍 NFA 的理论知识。

7.1 什么是复杂事件处理

7.1.1 股票异常交易检测

股市中每一笔交易的背后都暗含着交易者的某种动机或意图，异常行为往往隐藏在一系列交易数据上，因此监管部门通常采用数据分析的方法挖掘违规交易行为，例如，某只股票连续三个交易日的收盘价格涨幅偏离值累计超过 20%，就属于异常交易价格波动。

某只股票的涨跌幅偏离值是其当日涨跌幅减去其所在证券交易所标的指数（上海证券交易所为上证指数，深圳证券交易所为深证成指），如某股票今天涨了 10%，而指数只涨了 1%，那么这只股票今天的涨幅偏离值就是 9%；相反，某股票今天跌了 10%，而指数涨了 1%，那么今天这只股票的跌幅偏离值就是 11%。

为了高效地找出存在异常交易波动的股票，我们使用模式匹配语言 SASE+ 编写程序。

SASE+ 是一种支持 Kleene 闭包的复杂事件处理语言，应用于流处理的 Kleene 闭包模式，其特征有事件定义、事件选择和终止原则。这些特征使它区别于传统问题中的模式研究。

SASE+ 的语言结构如下：

```
[FROM <input stream>]
PATTERN <pattern structure>
[WHERE <pattern matching condition>]
[WITHIN <sliding window>]
[HAVING <pattern filtering condition>]
RETURN <output specification>
```

PATTERN 子句声明了模式结构；WHERE 子句定义事件谓词，类似 SQL 中的 where 子句；WITHIN 子句定义事件的时间窗口；HAVING 子句是定义在每个模式上的过滤事件；RETURN 子句输出模式匹配的事件结果。

我们用 id 表示股票代码，price 表示股票当日的收盘价格，volume 表示当日成交量，deviation 表示这只股票的涨跌幅偏离值，date 表示对应的日期：

```
case class Stock (id:String, price:Long, volume:Long, deviation:Double, date:Date)
```

则在**单只股票数据**（input.keyBy("id")）**上检测异常波动**的程序如下（为了后续引用方便，将此程序命名为"Query 1 程序"）：

257

```
PATTERN SEQ(Stock a, Stock b, Stock c)
WHERE skip_till_next_match(a,b,c){
   a.deviation + b.deviation + c.deviation > 20%
or a.deviation + b.deviation + c.deviation < -20%
}
WITHIN 3 days
```

程序以 PATTERN 开始，SEQ 表示股票收盘统计事件按照日期排序；a、b 和 c 为连续三天的收盘统计；WHERE 定义匹配条件；skip_till_next_match 表示要匹配的三个事件应该连续，且满足累计涨跌幅偏离值条件；WITHIN 表示事件时间窗口大小。

下面看一个更复杂的例子，我们从实时交易数据中抓取符合某种特征的股票，如在过去一小时内在某个时间点的成交量突然拉高，随后伴随着成交价格总体上的持续走高，成交量骤然下跌。输入是证券交易所的所有股票的实时交易数据，对应代码如下（将此程序命名为"Query 2 程序"）：

```
PATTERN SEQ(Stock+ a[], Stock b)
WHERE skip_till_next_match(a[],b){
   [id]
and a[1].volume > 1000
and a[i].price > avg(a[..i-1].price)
and b.volume < 80%*a[a.LEN].volume
}
WITHIN 1 hours
```

Stock+ 为 Kleene 闭包（L+，即匹配 L 一次或多次）；a[]表示将符合条件的交易记录存储在缓存 a 中；a[1]是初始交易记录，即交易量突然被拉高的事件，随后总体成交价格持续上升（a[i].price > avg(a[..i-1].price)），在交易记录 b 的时间点交易量下跌了 20%以上。其中，a[..i-1]表示 a 中的前 i-1 条记录；a.LEN 表示 a 中最后一条记录，a 的下标从 1 开始；[id]表示股票数据分区，与 DataStream API 中 keyBy("id")的语义相同。

7.1.2 重新审视 DataStream 与 Table API

对于流式数据处理，我们已经介绍了两类 API。

（1）DataStream API，是流式计算理论的分布式实现，它的核心思想是用有向无环计算图表示数据处理流程。但是，基于这种原生 API 实现复杂的模式匹配会过度依赖编程技巧，正如关系型 API 中分析的那样，降低应用门槛以便让数据分析人员专注于开发业务逻辑代表了先进架构的生产力。

（2）Table API 与 SQL，是基于动态表的、关系型数据库查询的 Flink 实现，是原生 API 的进一步封装，以降低应用程序编写门槛，其核心思想是"查询-连接-聚合流（select-join-aggregate）"，不具备复杂模式匹配基因。

因此，我们需要一种全新的、定义在流式架构上的模式匹配 API，如复杂谓词、多种匹配模式、闭包。

7.2 复杂事件处理的自动机理论

7.2.1 有穷自动机模型 NFA

从应用层面上我们很容易讲清楚什么是计算机，例如，可以说那台有触摸屏、能上网的设备就是计算机，但是从理论层面上建模却相当复杂。

为了探究计算机的基本能力和局限性，数理逻辑学家们从不同角度进行了深入研究，试图为计算机建立一个理想的、易于处理的数学模型，并在自动机、可计算性、复杂性领域分别给出了答案，而自动机是这些研究的基本工具。

按照对计算机刻画精准程度的不同，有不同的自动机模型，这其中最简单的是有穷自动机（Finite Automaton）模型。

图 7-1 描述了一台有三个状态的有穷自动机 M1 的状态图（State Diagram）。

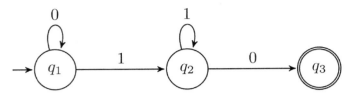

图 7-1 有穷自动机 M1

有穷自动机是为计算机建模而引入的,因此在有穷自动机 M1 中我们假定状态机的输入是由 0 和 1 组成的字符串。有穷自动机 M1 有三个状态,记作 q_1、q_2 和 q_3:用一个无出发点的箭头表示起始状态(如图 7-1 中起始状态 q_1 左侧的箭头所示);用双圈表示接受状态(如图 7-1 中的 q_3 所示);用从一个状态指向另一个状态的箭头表示转移,箭头上的数字表示状态转移的条件;这个机器处理接收的输入字符串,输出是接受或拒绝,即模式匹配中的匹配或不匹配。

当这个机器接收到字符串 110 时,状态的转移过程如下:

(1)初始状态为 q_1。

(2)读入第一个字符 1,状态转移到 q_2。

(3)读入字符 1,状态保持为 q_2。

(4)读入字符 0,状态转移到 q_3。

(5)最终状态为接受状态,因此输出为接受。

有穷自动机定义为一个五元组 $(Q, \Sigma, \delta, q_0, F)$,描述如下:

(1)称有穷集合 Q 为状态集。

(2)称有穷集合 Σ 为字母表,即允许的输入符号的集合。在有穷自动机 M1 中,字母表中仅有 0 和 1 两个字符。

(3)称 $\delta : Q \times \Sigma \rightarrow Q$ 为转移函数。转移函数描述了有穷自动机状态图中所有的状态转移规则。

（4）$q_0 \in Q$ 是初始状态。

（5）$F \subseteq Q$ 是接受状态集，有穷自动机允许有多个接受状态，也允许没有接受状态。

在**确定型**有穷自动机（DFA，Deterministic Finite Automaton）M1 中，我们发现每一个状态的所有转移条件是互斥的，或者说，自动机每次读入一个符号时可以事先知道机器的下一个状态，如状态q_1在转移条件为 0 时转移至q_1，在转移条件为 1 时转移至q_2。而在**非确定型**有穷自动机（NFA，Nondeterministic Finite Automaton）中，对于任何一个状态，要转移的下一个状态可能存在若干个非互斥的选择，图 7-2 所示为非确定型有穷自动机 M2。

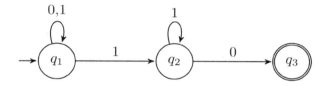

图 7-2　非确定型有穷自动机 M2

状态q_1对转移条件 1 有两个转移箭头，因此，当读入 1 时状态机分为两个过程，其中一个过程中的机器状态为q_2，另一个过程中的机器停留在q_1状态，当再次读入 1 时，一个过程继续分为两个过程。因此，可以将一台 NFA 拆解成一台 DFA，即每一台非确定型有穷自动机都等价于某一台确定型有穷自动机，但是由于构造 NFA 通常比直接构造 DFA 更容易，且 NFA 的功能更易于理解，在基于自动机的构造设计中通常采用 NFA 形式论证，如接下来要分析的 NFA[b] 模型。

7.2.2　NFA[b] 模型

事件处理 NFA[b] 模型包括一台 NFA 和一个匹配缓冲区（match buffer），表示为 $A = (Q, E, \theta, q_1, F)$，其中$Q$为状态集，$E$为有向边，$\theta$为状态转移条件，$q_1$为初始状态（为了和 SASE+语言保持一致，下标从 1 起始），F为接受状态。图 7-3 所示为 Query 2 程序对应的自动机 NFA[b] 模型（以下简称为 Query 2 自动机）。

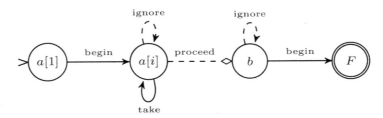

图 7-3 Query 2 程序对应的 NFA^b 模型

1. 状态

在 Query 2 自动机中，以缓存名称命名状态，以起始状态 $a[1]$ 为例，自动机将初始事件存储在 $a[1]$ 中，这是 Kleene 闭包的起始点。后续匹配的事件将存储在 $a[i]$ 中，对应状态为 $a[i]$。在 Kleene 闭包匹配完成后，b 存储下一个匹配事件。

2. 边

每一条边定义一种状态转移路径。有 4 类边：take 表示事件匹配成功，用于当前状态到自身的转移，如 Query 2 自动机中的 $a[i]$ 状态，take 将当前事件存储在 $a[i]$ 中；ignore 表示事件匹配不成功，当前状态不转移，与 take 的区别是当前事件不存储在 $a[i]$ 中；proceed 表示转移下一个状态，当前事件不存储在当前状态对应的缓存中；begin 表示当前状态的开始，在 Query 2 自动机中，状态 $a[1]$ 匹配成功后，下一个事件将进入 $a[i]$ 状态。

边有三个转移属性：第一个是转移条件，用状态加转移类型表示，如 $\theta_{a[i]\_take}$；第二个是对输入的操作，如消费当前事件（$\theta_{a[i]\_take}$ 匹配当前事件，再读入新的事件）；第三个是匹配缓存的操作，如将匹配的事件写入缓存。

无界数据集上的 NFA^b 模型是复杂的，因为无法定义自动机的匹配起点。因此，同一个窗口内的数据可能会对应多个并行的模式匹配过程。对于 Query 2 程序，给定如表 7-1 所示的事件输入。

表 7-1 Query 2 程序的事件输入

事件	e1	e2	e3	e4	e5	e6	e7	e8
price	100	120	120	121	120	125	120	120
volume	1010	990	1005	999	999	750	950	700

可能有多种匹配过程,如 R1、R2 和 R3。

表 7-2 Query 2 程序的三种匹配

匹配过程	状态 a[]	状态 b
R1	[e1, e2, e3, e4, e5]	e6
R2	[e3, e4]	e6
R3	[e1, e2, e3, e4, e5, e6, e7]	e8

此外,非确定型有穷自动机在执行过程中也会存在多个分叉过程。这种并行模式匹配过程不仅导致运行时异常复杂,也会拉长执行时间。

7.2.3 带版本号的共享缓存

对 Query 2 程序的三种匹配过程 R1、R2 和 R3 而言,要为每种匹配单独分配缓存(接受状态不需要缓存),如图 7-4 所示。

其中,每个匹配事件都有一个指针,即图 7-4 中的箭头,这类似数据结构中链表的指针,它指向当前元素的上一个匹配元素。

我们发现这些并存过程的独立缓存之间有重叠,因此从性能角度考虑,多个并行过程共享同一缓存可以避免缓存呈爆炸式增长。为了有效地管理共享缓存,NFA^b 模型为每一个并行过程分配一个版本号,并将此版本号附加在事件指针上。由于这些并行过程在结构上有关联,如 NFA 中的分叉,版本号可以设计成有继承关系的结构。基于这两点考虑,NFA^b 模型基于杜威十进制分类法(Dewey Decimal Classification)生成版本号。

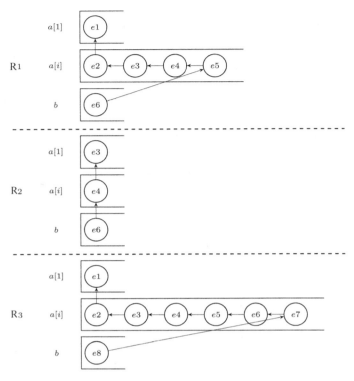

图 7-4 Query 2 程序的三种匹配过程 R1、R2 和 R3 的独立缓存

在 NFA[b] 模型的基本结构中,状态是串行推进的,因此可以按照 begin 和 proceed 的箭头方向将状态编号,如 q_t 表示第 t 个状态。这种版本号根据状态的推进动态增长,形式为

$$\mathrm{id}_1(.\mathrm{id}_j) \quad (1 \leqslant j \leqslant t)$$

其中 t 为当前状态 q_t,上述版本号表示状态机处于第 id_1 个过程,状态位置 q_j 处于第 id_j 个分叉。根据这种方法,共享缓存可绘制为如图 7-5 所示的样子。

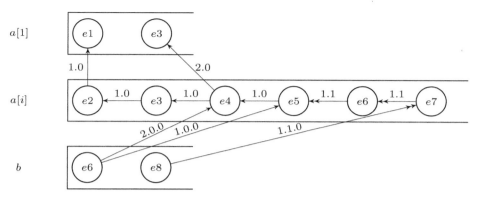

图 7-5 Query 2 程序的三种匹配过程 R1、R2 和 R3 的共享缓存

R1 是 R3 在事件 **e6** 上的同源分叉，所以将状态 $a[i]$ 上的事件缓存 **e6** 和 **e7** 的版本号加 1，即 1.1；R3 在状态 b 上的事件缓存 **e8** 则因为状态向前推进而增长为 1.1.0；R2 为另一个独立匹配过程，版本号的第一位为 2。

7.3 FlinkCEP API

计算图是 Flink 组织流处理节点的架构形式，将复杂事件处理抽象成模式图的形式是这种思路的延续。类比计算图中的节点，模式图由基本模式构建，并以拼合、分组的方式形成复杂的处理模式。

FlinkCEP（Complex Event Process for Flink）是以库的形式提供的，在 Maven 工程中需要加入以下依赖：

```
<properties>
    <flink.version>1.6.1</flink.version>
    <scala.binary.version>2.11</scala.binary.version>
</properties>

<dependency>
    <groupId>org.apache.flink</groupId>
    <artifactId>flink-cep-scala_${scala.binary.version}</artifactId>
    <version>${flink.version}</version>
    <scope>compile</scope>
```

```
</dependency>
```

首先定义模式,然后匹配输入,最后输出匹配结果,FlinkCEP 应用程序的例子如下:

```
val input: DataStream[Event] = ...

// 定义模式
val pattern = Pattern.begin[Event]("start").where(_.id == "42")
  .next("middle").subtype(classOf[SubEvent]).where(_.volume >= 1000)
  .followedBy("end").where(_.price == 125)

// 匹配输入
val patternStream = CEP.pattern(input, pattern)

// 输出匹配结果
val result: DataStream[Alert] = patternStream.select(createAlert(_))
```

其中,"start""middle""end"是应用程序为每个模式自定义的名称。

7.3.1 基本模式

基本模式(Individual Pattern)分为两类:单例模式(Singleton)和循环模式(Looping),其中单例模式只读入单个事件,而循环模式读入多个事件(如 Kleene 闭包 Stock+)。单例模式可以通过量词转化成循环模式,如 times、oneOrMore。对 pattern 而言,以下是转化成循环模式的方式:

```
// pattern 模式出现 4 次
pattern.times(4)

// pattern 模式出现 0 次或 4 次
pattern.times(4).optional()

// pattern 模式出现 2 次、3 次或 4 次
pattern.times(2, 4)

// pattern 模式出现 2 次或更多次
pattern.oneOrMore(2)
```

FlinkCEP 通过 where 定义状态转移条件，有以下 4 种类型：

（1）可迭代条件，除了需要向 filter 方法传入当前事件，还需要传入上下文，这样利用当前条件判断可以遍历之前匹配成功的事件。假定事件包括 price 属性，下例将之前匹配事件和当前事件的 price 累加求和，如果和没有超过 500，则当前事件进入自动机。

```
middle.oneOrMore()
  .subtype(classOf[SubEvent])
  .where(
    (value, ctx) => {
      lazy val sum = ctx.getEventsForPattern("middle").map(_.price).sum
      sum + value.price < 500
    }
  )
```

（2）简单条件，这类条件没有上下文参数。通过下例可以判断当前事件的 price 参数是否超过某个固定的值。

```
start.subtype(classOf[SubEvent]).where(subEvent => subEvent.price < 150)
```

（3）条件拼合，即前述两类条件的逻辑运算，如逻辑与（and）、逻辑或（or）。

```
pattern.where(event => ... /* some condition */).or(event => ... /* or condition */)
```

（4）停止条件，如单例模式转化成循环模式的 oneOrMore。

7.3.2　模式拼合

基本模式通过不同的连接方式可以组合成更复杂的模式，FlinkCEP 提供如下三种模式。

（1）严紧邻模式（Strict Contiguity），方法为 next() 或 notNext()，紧邻的两个模式必须按照先后顺序匹配，且都匹配。

```
val strict: Pattern[Event, _] = start.next("middle").where(...)
```

（2）松紧邻模式（Relaxed Contiguity），方法为 followedBy() 或 notFollowedBy()，

即忽略两个模式均不匹配的事件，从输入中找到符合这两个模式的事件流。

```
val relaxed: Pattern[Event, _] = start.followedBy("middle").where(...)
```

（3）非确定型松紧邻模式（Non-Deterministic Relaxed Contiguity），方法为 followedByAny()，这种模式可以忽略某些匹配事件。

```
val nonDetermin: Pattern[Event, _] = start.followedByAny("middle").where(...)
```

为了理解这三种模式的区别，我们举个简单的例子。假定输入为("a", "c", "b1", "b2")，其中的数字为第几个的意思，如"b2"代表出现的第二个事件"b"，以此区分匹配结果的位置。要匹配出"a b"的模式，即 a 的后面跟着一个 b：

（1）对于严紧邻模式，匹配的结果为空。

（2）对于松紧邻模式，匹配的结果为{a b1}，其语义为忽略不匹配的事件直到下一个匹配事件为止。

（3）对于非确定型松紧邻模式，匹配的结果为{a b1}和{a b2}

7.3.3 模式分组

FlinkCEP 引入模式分组机制，以提升模式编程的灵活度。所谓分组是将 begin、followedBy、followedByAny、next 的序列组合成一个基本模式，从而运用基本模式的量词：

```
val start: Pattern[Event, _] = Pattern.begin(
  Pattern.begin[Event]("start").where(...).followedBy("start_middle").where(...)
)

// strict contiguity
val strict: Pattern[Event, _] = start.next(

Pattern.begin[Event]("next_start").where(...).followedBy("next_middle").where(...)
).times(3)

// relaxed contiguity
val relaxed: Pattern[Event, _] = start.followedBy(
```

```
Pattern.begin[Event]("followedby_start").where(...).followedBy("followedby_middle").
where(...)
).oneOrMore()

// non-deterministic relaxed contiguity
val nonDetermin: Pattern[Event, _] = start.followedByAny(

Pattern.begin[Event]("followedbyany_start").where(...).followedBy("followedbyany_mid
dle").where(...)
).optional()
```

7.3.4 匹配输出

用模式编程的目的是输出匹配的输入序列,可以通过 select()或 flatSelect()输出结果。

(1) select()有多种原型。

```
def select[R: TypeInformation](patternSelectFunction: PatternSelectFunction[T, R])
: DataStream[R] = {

def select[L: TypeInformation, R: TypeInformation](
  outputTag: OutputTag[L],
  patternTimeoutFunction: PatternTimeoutFunction[T, L],
  patternSelectFunction: PatternSelectFunction[T, R])
: DataStream[R] = {

def select[R: TypeInformation](patternSelectFun: Map[String, Iterable[T]] => R):
DataStream[R] = {

def select[L: TypeInformation, R: TypeInformation](outputTag: OutputTag[L])(
  patternTimeoutFunction: (Map[String, Iterable[T]], Long) => L) (
  patternSelectFunction: Map[String, Iterable[T]] => R)
: DataStream[R] = {
```

这些形式的核心参数是 patternSelectFun。Map[String, Iterable[T]]中的第一个参数为 key,是模式的名称,如"start"等;第二个参数为每种模式的事件匹配结果。下面的编程实例将"start"和"end"模式的匹配结果输出(OUT(…))。

```
def selectFn(pattern : Map[String, Iterable[IN]]): OUT = {
  val startEvent = pattern.get("start").get.next
  val endEvent = pattern.get("end").get.next
  OUT(startEvent, endEvent)
}
```

最后，我们可以将以上函数传入 select() 方法中得到想要的结果。

（2）相应地，flatSelect() 也有多种原型，patternFlatSelectFunction 的输出是一个 Collector，它与 select() 的区别类似 flatMap 和 map 的区别。

```
def flatSelect[R: TypeInformation](patternFlatSelectFunction:
PatternFlatSelectFunction[T, R])
: DataStream[R] = {

def flatSelect[L: TypeInformation, R: TypeInformation](
  outputTag: OutputTag[L],
  patternFlatTimeoutFunction: PatternFlatTimeoutFunction[T, L],
  patternFlatSelectFunction: PatternFlatSelectFunction[T, R])
: DataStream[R] = {

def flatSelect[R: TypeInformation](patternFlatSelectFun: (Map[String, Iterable[T]],
  Collector[R]) => Unit): DataStream[R] = {

def flatSelect[L: TypeInformation, R: TypeInformation](outputTag: OutputTag[L])(
  patternFlatTimeoutFunction: (Map[String, Iterable[T]], Long, Collector[L]) => Unit)
(
  patternFlatSelectFunction: (Map[String, Iterable[T]], Collector[R]) => Unit)
: DataStream[R] = {
```

7.4 基于 FlinkCEP 的股票异常交易检测的实现

以表 7.1 作为输入事件，用 FlinkCEP 实现 Query 2 程序的功能。首先，创建 DataStream[Stock]。

```
val input: DataStream[Stock] = env.fromElements(
  "100,1010",
  "120,990",
  "120,1005",
```

```
    "121,999",
    "120,999",
    "125,750",
    "120,950",
    "120,700"
).map({
  x =>
    val y: Array[String] = x.split(",")
    Stock(y(0).trim().toInt, y(1).trim().toInt)
}).setParallelism(1)
```

然后定义模式,起始点为成交量高于 1000 的事件。

```
val pattern = Pattern.begin[Stock]("start").where(_.volume > 1000)
```

紧接着,由可迭代条件定义模式,将严紧邻模式和起始模式拼合,并由量词(oneOrMore)将单例模式转化成循环模式。利用下面的模式求已匹配的所有事件收盘价的平均值,并判断当前读入事件的收盘价是否大于平均值:

```
.next("middle").subtype(classOf[Stock]).where(
    (value, ctx) => {
      val startSum = ctx.getEventsForPattern("start").map(_.price).sum
      val count    = ctx.getEventsForPattern("middle").size
      if (count > 0) {
        val sum  = ctx.getEventsForPattern("middle").map(_.price).sum
        value.price > (sum + startSum)/(count + 1.0)
      }
      else
        value.price > startSum
    }
).oneOrMore
```

最后是接受状态,即当前读入事件的成交量小于已匹配的最后一个事件的成交量的 80%:

```
.followedBy("end").subtype(classOf[Stock]).where(
    (value, ctx) => {
      val count       = ctx.getEventsForPattern("middle").size
      if (count > 0) {
        val stock: Stock = ctx.getEventsForPattern("middle").toList.apply(count - 1)
        value.volume < 0.8 * stock.volume
      }
```

```
      else
        false
    }
  )

val patternStream: PatternStream[Stock] = CEP.pattern(input, pattern)
```

定义完模式后，定义匹配输出的 select 方法：

```
val result: DataStream[List[Stock]] = patternStream
  .select((pat: Map[String, Iterable[Stock]]) => {
    val startEvent: Stock           = pat.get("start").get.head
    val middleEvent: Iterator[Stock]  = pat.get("middle").get.iterator
    val endEvent: Stock             = pat.get("end").get.head
    val start: List[Stock]          = List(startEvent)
    val middle = start ::: middleEvent.toList
    val out    = middle ::: List(endEvent)
    out
  })
```

上述程序的输出如下：

```
List(100_1010, 120_990, 120_1005, 121_999, 120_999, 125_750)
List(100_1010, 120_990, 120_1005, 121_999, 125_750)
List(100_1010, 120_990, 120_1005, 125_750)
List(100_1010, 120_990, 125_750)
List(120_1005, 121_999, 125_750)
List(100_1010, 120_990, 120_1005, 121_999, 120_999, 125_750, 120_950, 120_700)
```

其中，第 1 行、第 5 行和第 6 行分别为 Query 2 程序的 3 种匹配过程 R1、R2 和 R3。

为便于整体阅读，所有代码如下：

```
package com.deepmind.flink

import org.apache.flink.cep.scala.{CEP, PatternStream}
import org.apache.flink.cep.scala.pattern.Pattern
import org.apache.flink.streaming.api.scala.{DataStream, StreamExecutionEnvironment,
  _}

import scala.collection.Map
```

```scala
object FlinkCEPSelect {
  // 定义 Stock 类
  case class Stock(price:Int, volume:Int){
    override def toString: String = price.toString + "_" + volume.toString
  }
  def main(args: Array[String]) {
    val env = StreamExecutionEnvironment.getExecutionEnvironment
    // 根据字符串类型的输入创建 DataStream[Stock]
    val input: DataStream[Stock] = env.fromElements(
      "100,1010",
      "120,990",
      "120,1005",
      "121,999",
      "120,999",
      "125,750",
      "120,950",
      "120,700"
    ).map({
      x =>
        val y: Array[String] = x.split(",")
        Stock(y(0).trim().toInt, y(1).trim().toInt)
    }).setParallelism(1)
    // 定义模式
    val pattern = Pattern.begin[Stock]("start").where(_.volume > 1000)
      .next("middle").subtype(classOf[Stock]).where(
        (value, ctx) => {
          val startSum = ctx.getEventsForPattern("start").map(_.price).sum
          val count    = ctx.getEventsForPattern("middle").size
          if (count > 0) {
            val sum  = ctx.getEventsForPattern("middle").map(_.price).sum
            value.price > (sum + startSum)/(count + 1.0)
          }
          else
            value.price > startSum
        }
      ).oneOrMore
      .followedBy("end").subtype(classOf[Stock]).where(
        (value, ctx) => {
          val count    = ctx.getEventsForPattern("middle").size
          if (count > 0) {
            val stock: Stock = ctx.getEventsForPattern("middle").toList.apply(count - 1)
            value.volume < 0.8 * stock.volume
          }
          else
```

```
        false
      }
    )

  val patternStream: PatternStream[Stock] = CEP.pattern(input, pattern)
  // 定义patternSelectFun: 将start、middle、end缓存中的事件拼接在一起输出
  val result: DataStream[List[Stock]] = patternStream
    .select((pat: Map[String, Iterable[Stock]]) => {
      val startEvent: Stock              = pat.get("start").get.head
      val middleEvent: Iterator[Stock]   = pat.get("middle").get.iterator
      val endEvent: Stock                = pat.get("end").get.head
      val start: List[Stock]             = List(startEvent)
      val middle = start ::: middleEvent.toList
      val out    = middle ::: List(endEvent)
      out
    })

  //println(patternStream)
  result.print().setParallelism(1)

  env.execute("DynamicTable")
 }
}
```

7.5　思考题

（1）带版本号的缓存共享是为了提升内存使用效率，但这给架构增加了难度。这种内存使用效率能够有多大提升呢？

（2）自动机在执行过程中会存在多个分叉过程，因此 NFA^b 模型是否存在死锁问题？

（3）NFA^b 模型的表达能力如何？换句话说，NFA^b 模型是否能解决所有模式匹配的问题？

（4）FlinkCEP 增加了窗口生存期，如何进行垃圾回收呢？

（5）相比 SQL 语言，FlinkCEP 模式匹配语言的语义丰富性体现在哪里？

第 8 章 监控与部署

本章介绍与运维相关的三个主题：监控、部署及访问安全。8.1 节介绍指标的度量方式及监控接口；8.2 节介绍集群的部署模式，如 Standalone 集群模式、YARN 集群模式，以及高可用部署模式；8.3 节介绍如何确保外部访问 Flink 内部资源及 Flink 各节点时的数据传输安全。

8.1 监控

8.1.1 度量指标

监控是分布式系统不可缺少的组成部分，Flink 也不例外。这里提及的监控与日志的区别是前者是动态的（准）实时系统监控方法，如系统级监控（CPU、内存、线程、I/O、垃圾回收、网络）、应用级监控（可用性、检查点、连接器、存储后端压力、事件时间与窗口、延迟），而后者是事后静态分析方法。

监控分为两部分，即 Metrics 和 Reporter。Metrics 用于度量指标；Reporter 用于向监控展示工具实时上传指标。我们可以通过 Flink 的 Dashboard 观测集群上所有应用程序定义的指标，也可以通过 Flink 提供的 REST API 获取这些指标。

应用程序通过 RichFunction 定义每一种指标，通过名字访问对应指标。定义指标有 4 种方式。

（1）计数器（Counter）。计数器用于统计总量，可增可减。

```
var counter: Counter = ...
// 数值加1
counter.inc()
// 数值加n
counter.inc(n)
// 数值减1
counter.dec()
// 数值减n
counter.dec(n)
```

以下例子实现了输入计数，并将输入字符串映射成小写格式，在 open 方法中创建并注册名字为 myCounter 的计数器，在 map 方法中操纵计数器。变量 counter 是计数器的瞬时值，不参与序列化，因此加上了修饰符@transient：

```
class MyMapper extends RichMapFunction[String,String] {
  @transient private var counter: Counter = _
  // 在 map 函数初始化时，定义名字为 myCounter 的 Counter
  override def open(parameters: Configuration): Unit = {
    counter = getRuntimeContext()
      .getMetricGroup()
      .counter("myCounter")
  }
  override def map(value: String): String = {
    // 计算器加1
    counter.inc()
    // 将字符串转换成小写格式
    value.toLowerCase
  }
}
```

应用程序也可以提供 Counter 的实现，以下是其 Java 语言的原型：

```
public interface Counter extends Metric {
    void inc();
    void inc(long n);
    void dec();
```

```
    void dec(long n);
    long getCount();
}
```

在 open 方法中传入 Counter 的对应实现：

```
class MyMapper extends RichMapFunction[String,String] {
  @transient private var counter: Counter = _
  override def open(parameters: Configuration): Unit = {
    counter = getRuntimeContext()
      .getMetricGroup()
      // 应用实现的 Counter
      .counter("myCustomCounter", new CustomCounter())
  }
  override def map(value: String): String = {
    counter.inc()
    value.toLowerCase
  }
}
```

（2）度量器（Gauge）。计数器是数值型指标，而度量器并不限定指标的数据类型。此外，没有内置的 Gauge 实现，应用程序需要自己实现 Gauge，以下是 Gauge 的 Scala 版本：

```
// 定义 Gauge 的 get 方法
class ScalaGauge[T](func: () => T) extends Gauge[T] {
  override def getValue: T = {
    func()
  }
}
```

以下程序实现了输入计数：

```
class MyMapper extends RichMapFunction[String,String] {
  @transient private var valueToExpose = 0
  override def open(parameters: Configuration): Unit = {
    getRuntimeContext()
      .getMetricGroup()
      // get 方法直接获取 valueToExpose 值
      .gauge[Int, ScalaGauge[Int]]("MyGauge", ScalaGauge[Int]( () => valueToExpose ) )
  }
  override def map(value: String): String = {
```

```
  valueToExpose += 1
  value.toLowerCase
 }
}
```

（3）直方图（Histogram）。直方图代表一个观察窗口而不是单值对象，主要用来统计数据的分布情况，如最大值、最小值、平均值、中位数、百分比等。Flink 没有提供默认实现，而是封装了 Codahale/DropWizard 的 Histogram 类：

```
class MyMapper extends RichMapFunction[Long, Long] {
 @transient private var histogram: Histogram = _
 override def open(config: Configuration): Unit = {
  // 定义直方图窗口
  com.codahale.metrics.Histogram dropwizardHistogram =
    new com.codahale.metrics.Histogram(new SlidingWindowReservoir(500))

  histogram = getRuntimeContext()
    .getMetricGroup()
    .histogram("myHistogram", new DropwizardHistogramWrapper(dropwizardHistogram))
 }
 override def map(value: Long): Long = {
  // 更新 value 值
  histogram.update(value)
  value.toLowerCase
 }
}
```

（4）计量器（Meter）。计量器用来度量吞吐量。类似 Histogram，Flink 没有提供默认实现，而是封装了 Codahale/DropWizard 的 Metrics 类。其中，markEvent() 接口用于标记事件的出现：

```
class MyMapper extends RichMapFunction[Long,Long] {
 @transient private var meter: Meter = _
 override def open(config: Configuration): Unit = {
  com.codahale.metrics.Meter dropwizardMeter = new com.codahale.metrics.Meter()
  meter = getRuntimeContext()
    .getMetricGroup()
    .meter("myMeter", new DropwizardMeterWrapper(dropwizardMeter))
 }
 override def map(value: Long): Long = {
  // 标记事件的出现
```

```
  meter.markEvent()
  value.toLowerCase
  }
}
```

8.1.2 指标的作用域

每一个指标绑定一个作用域，有用户自定义作用域（user-defined scope）和系统级作用域（system-provided scope）两种作用域。每一种作用域是一组键值对：

（1）用户自定义作用域。可以通过以下三种方式定义。

- MetricGroup.addGroup(String name)
- MetricGroup.addGroup(int name)
- Metric.addGroup(String key, String value)

（2）系统级作用域。定义指标的物理路径，host 表示任务所在主机；job_name 表示作业名称；task_name 表示任务名称；operator_name 表示算子名称；subtask_index 表示是第几个并行实例，指标的作用域形式如下所示：

```
<host>.<job_name>.<task_name>.<operator_name>.<subtask_index>
```

8.1.3 监控配置

Flink 通过自己的调度程序（准）实时地推送指定作用域内的指标度量结果给外部监控系统，而 Reporter 则可以看作外部监控系统的客户端。Flink 提供以下几类 Reporter，需要在配置文件中配置。

1. JMX

```
// Reporter 类型
metrics.reporter.jmx.class: org.apache.flink.metrics.jmx.JMXReporter
// Flink 监听端口号
metrics.reporter.jmx.port: 8789
```

2. Graphite

```
// Reporter 类型
metrics.reporter.grph.class: org.apache.flink.metrics.graphite.GraphiteReporter
```

```
// Graphite 服务端 IP
metrics.reporter.grph.host: localhost
// Graphite 服务端端口号
metrics.reporter.grph.port: 2003
// Graphite 服务端协议
metrics.reporter.grph.protocol: TCP
```

3. Prometheus

```
// Reporter 类型
metrics.reporter.prom.class: org.apache.flink.metrics.prometheus.PrometheusReporter
// Flink 监听端口号
metrics.reporter.prom.port: 8249
```

4. StatsD

```
// Reporter 类型
metrics.reporter.stsd.class: org.apache.flink.metrics.statsd.StatsDReporter
// StatsD 服务端 IP
metrics.reporter.stsd.host: localhost
// StatsD 服务端端口号
metrics.reporter.stsd.port: 8125
```

5. Datadog

```
// Reporter 类型
metrics.reporter.dghttp.class: org.apache.flink.metrics.datadog.DatadogHttpReporter
// API key
metrics.reporter.dghttp.apikey: xxx
// API tag, 可选
metrics.reporter.dghttp.tags: myflinkapp,prod
```

6. Slf4j

```
// Reporter 类型
metrics.reporter.slf4j.class: org.apache.flink.metrics.slf4j.Slf4jReporter
// 文件切分的时间间隔
metrics.reporter.slf4j.interval: 60 SECONDS
```

此外，Flink 还提供应用自定义的 Reporter 接口，这么做的代价是应用程序需实现调度接口。

8.2 集群部署模式

8.2.1 Standalone

Flink 集群运行在类 UNIX 操作系统上，采用 master/slave 架构，因此至少包括一个 master 节点和一个 worker 节点，主要部署步骤如下：

（1）安装 Java 8（或更高版本）。

（2）配置 SSH 免密码登录。master 通过操作系统 SSHD 服务登录 slave 执行管理命令，以下是 master 通过 SSH 启动 worker 节点的命令（摘自 config.sh）：

```
ssh -n $FLINK_SSH_OPTS $slave -- "nohup /bin/bash -l \"${FLINK_BIN_DIR}/taskmanager.sh\" \"${CMD}\" &"
```

因此，master 需要免密码登录所有 worker 节点。

（3）配置集群结构，即配置 conf/masters 和 conf/slaves。此外，还需要确保 master 和 slave 的 Flink 部署目录结构一致。

（4）启动集群。在 master 上执行以下命令：

```
/bin/flink run WordCount.jar --para "value"
```

8.2.2 YARN

YARN 的出现与 Hadoop 的成功密不可分。在 Hadoop1.0 版中，NameNode 和 JobTracker 绑定在一个节点上，不仅会导致单点故障，也阻碍了 Hadoop 集群扩展规模：

（1）在 Hadoop 1.0 版中，JobTracker 要同时完成资源管理和作业控制，仅作业控制和资源管理的心跳信息就会让 JobTracker 难以应付。

（2）Hadoop 1.0 版的架构满足不了除 Hadoop MapReduce 外的其他计算框架的集成需求，因此功能拆分是有效的解决方案。

YARN 将 JobTracker 的资源管理和作业控制拆成如下几个部分（Daemon）：

（1）全局的资源管理器（RM，ResourceManager）。

（2）作业控制器（AM，ApplicationMaster），每一个作业对应一个作业控制器。

（3）节点管理器（NodeManager），每一个节点（DataNode）对应一个节点管理器。

YARN 架构如图 8-1 所示，其中 App、Mser 是作业控制器。

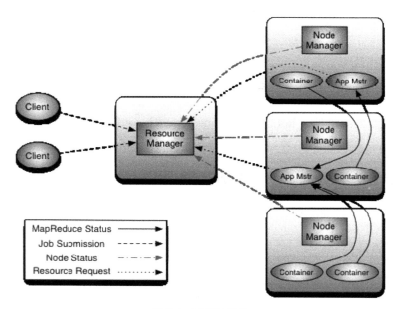

图 8-1　YARN 架构

Flink 的每一个版本均**支持**多个 Hadoop 版本号的发行版本，例如：

- Flink 1.6.2 对应 Hadoop® 2.8；
- Flink 1.6.2 对应 Hadoop® 2.7；
- Flink 1.6.2 对应 Hadoop® 2.6；
- Flink 1.6.2 对应 Hadoop® 2.4。

因此，我们必须将 Flink 集成到对应版本的 Hadoop 集群中。

Flink 集群的启动过程如下：

（1）创建 YARN session，在 Flink master 上运行以下命令：

```
./bin/yarn-session.sh -n 4 -jm 1024m -tm 4096m
```

此时，Flink 的 YARN Client 首先检测 YARN 和 HDFS 的配置，然后上传 Flink Jar 包和相关配置信息到 HDFS。其中，YARN Client 被 Flink 编译进安装包。

（2）Client 向 YARN 集群申请 AM 容器以启动 Flink 的 JobManager。

（3）JobManager 向 AM 申请容器以启动 TaskManager。

Flink 集群的启动过程如图 8-2 所示。

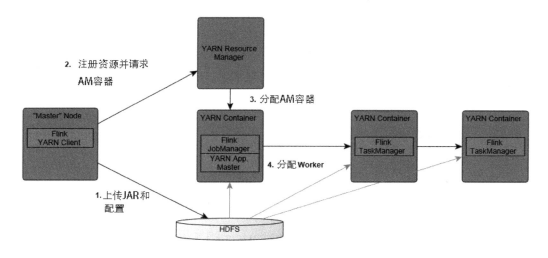

图 8-2　Flink 集群的启动过程

Flink 集群启动成功后，通过以下命令提交作业任务：

```
./bin/flink run -m yarn-cluster -yn 4 -yjm 1024m -ytm 4096m ./examples/batch/WordCount.jar
```

8.2.3 高可用

Flink 采用主从架构，JobManager 运行在 master 节点上，TaskManager 运行在 slave 节点上。主节点进程负责监控从节点状态并给从节点分配任务，这里可能存在以下三种异常情况：

（1）主节点崩溃。

（2）从节点崩溃。

（3）主从节点之间通信故障。

第一种异常是关键问题，后两种异常通常由主节点处理，所以这里定义的高可用指的是 JobManager 的高可用，部署多个 JobManager 是这个问题的解决方案。Flink 借助分布式过程协同组件 ZooKeeper 确保每个时间点只能有一个 JobManager 成为集群的 Leader，其他 JobManager 节点为 Standby。

而托管于 YARN 的 Flink 集群则不需要部署多个 JobManager 及额外的 ZooKeeper，YARN 负责保证 Flink 集群的高可用。

8.3 访问安全

Flink 通过 SSL（安全套接层，Secure Sockets Layer）和 TLS（传输层安全，Transport Layer Security）确保网络通信安全与数据完整性，其中访问安全包括：

（1）Flink 节点间通信安全，如 JobManager 和 TaskManager。

（2）外部系统访问内部的数据传输安全，如 WebUI（Dashboard）、命令行（CLI）、REST 接口。

为了确保架构的灵活性，Flink 将访问安全分为内部通信安全和外部访问安全两部分，如图 8-3 所示。

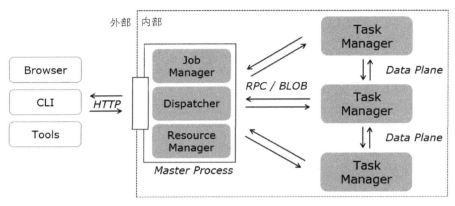

图 8-3 Flink 的访问安全

图 8-3 中的内部网络连接包括：

（1）多个组件间（如 JobManager、TaskManager、Dispatcher、ResouceManager 进程之间）通过 RPC 传递控制消息。

（2）TaskManager 间的数据分发。

（3）BLOB 服务和 RPC 请求监听服务。

其中，内部连接都是双向的，因此需要安全套接层互相授信，这通过配置 keystore 和 truststore 实现：

```
security.ssl.internal.keystore: /path/to/file.keystore
security.ssl.internal.keystore-password: keystore_password
security.ssl.internal.key-password: key_password
security.ssl.internal.truststore: /path/to/file.truststore
security.ssl.internal.truststore-password: truststore_password
```

此外，还需打开 Flink 内部的 SSL 机制：

```
security.ssl.internal.enabled:
| taskmanager.data.ssl.enabled
| blob.service.ssl.enabled
| akka.ssl.enabled
```

外部网络连接包括两部分：

（1）WebUI（或 CLI）和 Dispatcher 连接，如提交任务。

（2）WebUI（或 CLI）和 JobManager 连接，如任务控制。

对应的 SSL 开关如下所示：

```
security.ssl.rest.enabled
```

SSL 授信配置如下所示：

```
security.ssl.rest.keystore: /path/to/file.keystore
security.ssl.rest.keystore-password: keystore_password
security.ssl.rest.key-password: key_password
security.ssl.rest.truststore: /path/to/file.truststore
security.ssl.rest.truststore-password: truststore_password
security.ssl.rest.authentication-enabled: false
```

8.4 思考题

（1）为什么 Counter 要定义减法操作？在流式计算场景下有什么案例会用到减法操作？

参考资料

[1] Streaming 101: The world beyond batch, Tyler Akidau, August 5, 2015, https://www.oreilly.com/ideas/the-world-beyond-batch-streaming-101

[2] Streaming 102: The world beyond batch, Tyler Akidau, January 20, 2016, https://www.oreilly.com/ideas/the-world-beyond-batch-streaming-102

[3] The Evolution of Massive-Scale Data Processing, Tyler Akidau, Strata + Hadoop World London 2016

[4] Tyler Akidau, Robert Bradshaw, Craig Chambers, et al. The Dataflow Model: A Practical Approach to Balancing Correctness, Latency, and Cost in MassiveScale, Unbounded, OutofOrder Data Processing, http://www.cse.ust.hk/~weiwa/teaching/Fall16-COMP6611B/reading_list/Dataflow.pdf

[5] How to beat the CAP theorem, http://nathanmarz.com/blog/how-to-beat-the-cap-theorem.html, Thursday, October 13, 2011

[6] Jeffrey Dean, Sanjay Ghemawat. MapReduce: Simplified Data Processing on Large Clusters, http://research.google.com/archive/mapreduce-osdi04.pdf

[7] https://ci.apache.org/projects/flink/flink-docs-release-1.6/

[8] Paris Carbone, Gyula Fóra, Stephan Ewen, et al. Lightweight Asynchronous Snapshots for Distributed Dataflows, https://arxiv.org/ abs/1506.08603

[9] Sanjeev Kulkarni, Nikunj Bhagat, Maosong Fu, et al. Twitter Heron: Stream Processing at Scale, https://dl.acm.org/citation.cfm? doid=2723372.2742788

[10] History of Apache Storm and lessons learned, http://nathanmarz.com/blog/history-of-apache-storm-and-lessons-learned.html

[11] Craig Chambers, Ashish Raniwala, Frances Perry, et al. FlumeJava: Easy, Efficient Data-Parallel Pipelines, http://x86.cs.duke.edu/courses/fall13/cps296.4/838-CloudPapers/FlumeJava.pdf

[12] Matei Zaharia, An Architecture for Fast and General Data Processing on Large Clusters, https://cs.stanford.edu/~matei/application/cv.pdf

[13] K. Mani Chandy, Leslie Lamport, Distributed Snapshots: Determining Global States of Distributed Systems, http://www.cs.utexas.edu/users/lorenzo/corsi/cs380d/papers/p63-chandy.pdf

[14] Martin Kleppmann, A Critique of the CAP Theorem, https://arxiv.org/abs/1509.05393.

[15] 骆志刚，丁凡，蒋晓舟等. 复杂网络社团发现算法研究新进展，国防科技大学学报第 33 卷，第一期，文章编号：1001- 2486(2011) 01- 0047- 06

[16] 吕天阳，谢文艳，郑纬民等. 加权复杂网络社团的评价指标及其发现算法分析，物理学报 Acta Phys. Sin. Vol. 61, No. 21 (2012) 210511

[17] Jagrati Agrawal, Yanlei Diao, Daniel Gyllstrom, et al. Efficient Pattern Matching over Event Streams, https://people.cs.umass.edu/~immerman/pub/sase+sigmod08.pdf

[18] Fusheng Wang, Peiya Liu. Temporal Management of RFID Data, http://archive.systems.ethz.ch/www.dbis.ethz.ch/education/ws0708/infsyst_lab/rfid/rfid_resources/vldb05.pdf

[19] 林杨，市场微结构的股市交易异常行为检测，福建师范大学学报（自然科学版）Vol. 29 No. 1 Jan. 2013, 文章编号: 1000-5277(2013) 01-0031-05

[20] Lars Buitinck, Gilles Louppe, Mathieu Blondel, et al. API design for machine learning software:experiences from the scikit-learn project, https://arxiv.org/pdf/1309.0238.pdf

[21] Shai Shalev-Shwartz, Tong Zhang. Stochastic Dual Coordinate Ascent Methods for Regularized Loss Minimization, arXiv:1209.1873v2 [stat.ML] 30 Jan 2013

[22] D. Sculley, Gary Holt, Daniel Golovin, et al.Machine Learning:The High-Interest Credit Card of Technical Debt, https://news.ycombinator.com/item?id=8775772

[23] Martin Jaggi, Virginia Smith, Martin Takáč, et al. Communication-Efficient Distributed Dual Coordinate Ascent, https://arxiv.org/pdf/1409.1458.pdf

[24] Zeynep Batmaz, Ali Yurekli, Alper Bilge, et al. A review on deep learning for recommender systems:challenges and remedies, https://link.springer.com/article/10.1007%2Fs10462-018-9654-y

[25] S. Boyd, L. Vandenberghe, Convex Optimization, Cambridge University Press, 2004

[26] Cortes, C., Vapnik, V. (1995) Support-Vector Networks. Machine Learning, 20, 273-297

[27] John C. Platt, Sequential Minimal Optimization: A Fast Algorithm for Training Support Vector Machines, Technical Report MSR-TR-98-14, Microsoft Research

[28] D. J. Abadi, et al. Aurora: A New Model and Architecture for Data Stream Management. The VLDB Journal, 12(2):120-139, Aug. 2003

[29] T. Akidau, et al, MillWheel: Fault-Tolerant Stream Processing at Internet Scale. In Proc. of the 39th Int. Conf. on Very Large Data Bases (VLDB), 2013

[30] A. Alexandrov, et al. The Stratosphere Platform for Big Data Analytics. The VLDB Journal,23(6):939-964, 2014

[31] C. Chambers, et al. FlumeJava: Easy, E_cient Data-Parallel Pipelines. In Proc. of the 2010 ACM SIGPLAN Conf. on Programming Language Design and Implementation (PLDI), pages 363-375, 2010

[32] O. Boykin, et al. Summingbird: A Framework for Integrating Batch and Online MapReduce Computations. Proc.VLDB Endow., 7(13):1441-1451,Aug. 2014

[33] J. Dean, S. Ghemawat. MapReduce: Simpli_ed Data Processing on Large Clusters. In Proc. of the Sixth Symposium on Operating System Design and Implementation (OSDI), 2004

[34] J. Li, et al. Semantics and Evaluation Techniques for Window Aggregates in Data Streams. In Proceedings of the ACM SIGMOD Int. Conf. on Management of Data (SIGMOD), pages 311-322, 2005

[35] J. Li, et al. Out-of-order Processing: A New Architecture for High-performance Stream Systems. Proc. VLDB Endow., 1(1):274-288, Aug. 2008

[36] U. Srivastava, J. Widom. Flexible Time Management in Data Stream Systems. In Proc. of the 23rd ACM SIGMOD-SIGACT-SIGART Symp. on Princ. of Database Systems, pages 263-274, 2004

[37] M. Zaharia, et al. Resilient Distributed Datasets: A Fault-Tolerant Abstraction for In-Memory Cluster Computing. In Proc. of the 9th USENIX Conf. on Networked Systems Design and Implementation (NSDI), pages 15-28, 2012

[38] M. Zaharia, et al. Discretized Streams: Fault-Tolerant Streaming Computation at Scale. In Proc. of the 24th ACM Symp. on Operating Systems Principles, 2013

[39] A. Alexandrov, R. Bergmann, S. Ewen, et al. The stratosphere platform for big

dataanalytics. The VLDB JournalThe International Journal on Very Large Data Bases, 23(6):939–964, 2014

[40] R. Castro Fernandez, M. Migliavacca, E. Kalyvianaki, et al. Integrating scale out and fault tolerance in stream processing using operator state management. In Proceedings of the 2013 ACM SIGMOD international conference on Management of data, pages 725–736. ACM, 2013

[41] K. M. Chandy, L. Lamport. Distributed snapshots:determining global states of distributed systems. ACM Transactions on Computer Systems (TOCS), 3(1):63–75, 1985

[42] G. Tel. Introduction to distributed algorithms. Cambridge university press, 2000

[43] M. Zaharia, T. Das, H. Li, et al.Discretized streams: an efficient and fault-tolerant model for stream processing on large clusters. In Proceedings of the 4th USENIX conference on Hot Topics in Cloud Ccomputing, pages 10–10. USENIX Association,2012

[44] D. G. Murray, F. McSherry, R. Isaacs, et al. Naiad: a timely dataflow system.In Proceedings of the Twenty-Fourth ACM Symposium on Operating Systems Principles, pages 439–455.ACM, 2013

[45] Y. Low, D. Bickson, J. Gonzalez, et al. Distributed graphlab: a framework for machine learning and data mining in the cloud. Proceedings of the VLDB Endowment, 5(8):716–727, 2012

[46] J. Dean. Experiences with MapReduce, an abstraction for large-scale computation. In Parallel Architectures and Compilation Techniques (PACT), 2006

[47] Nicolas Liochon. You do it too: Forfeiting network partition tolerance in distributed systems, July 2015. URL http://blog.thislongrun.com/2015/07/Forfeit-

Partition-Tolerance-Distributed-System-CAP-Theorem.html

[48] I. Wegener. Branching programs and binary decision diagrams: theory and applications. Society for Industrial and Applied Mathematics, Philadelphia,PA, USA, 2000

[49] R. Sadri, C. Zaniolo, et al. Expressing and optimizing sequence queries in database systems. ACM Trans. Database Syst., 29(2):282-318, 2004

[50] F. Yu, Z. Chen, et al. Fast and memory-e_cient regular expression matching for deep packet inspection. In ANCS, 93-102, 2006

[51] D. Zimmer, R. Unland. On the semantics of complex events in active database management systems. In ICDE, 392-399, 1999

[52] S. Gatziu, K. R. Dittrich. Events in an active object-oriented database system. In Rules in Database Systems, 23-39, 1993

[53] S. Chandrasekaran, O. Cooper, et al. TelegraphCQ: Continuous dataow processing for an uncertain world. In CIDR, 2003